SMITH & WESSON

. . . .

The Story of the Revolver

. . . .

INCLUDES

SMITH & WESSON REVOLVERS AND THEIR CURRENT PRICES

By MARTIN RYWELL

. . . AND . . .

100 YEARS OF GUNMAKING

By CARL R. HELLSTROM

President, Smith & Wesson

PIONEER PRESS • Harriman, Tennessee

OTHER BOOKS BY MARTIN RYWELL

AMERICAN ANTIQUE RIFLES AND THEIR CURRENT PRICES
AMERICAN ANTIQUE GUNS AND THEIR CURRENT PRICES
U. S. MILITARY MUSKETS, RIFLES, CARBINES AND THEIR CURRENT PRICES
SAMUEL COLT—A MAN AND AN EPOCH
COLT GUNS—STORY OF COLT REVOLVER
CONFEDERATE GUNS AND THEIR CURRENT PRICES
JUDAH P. BENJAMIN, A STUDY OF CONFEDERATE DIPLOMACY
MORTAR IN AMERICAN FOUNDATIONS
DECIMAL INDEXES
INVESTIGATORS LAW MANUAL
ANTHOLOGY OF SHORT STORIES
TENNESSEE COOK BOOK
MEXICAN COOK BOOK
WILD GAME COOK BOOK

FIRST EDITION
COPYRIGHT, 1953 BY MARTIN RYWELL

All Rights Reserved

A PATHFINDER BOOKS REPRINT EDITION
Printed in the United States of America
ISBN: 978-1-951682-42-2

CONTENTS

Chapter	Page
1. Daniel Baird Wesson	5
2. Horace Smith	11
3. The Patent Background	13
4. Smith & Wesson	16
5. Volcanic Repeating Arms Co.	20
6. Smith & Wesson Again	23
7. The Rim-Fire Cartridge	27
8. 1861-1874	33
9. 1873—Smith Retired	41
10. 1875-1880	43
11. 1881-1906	47
12. Daniel Baird Wesson Died	53
13. 1906-1953	55
14. The Epoch	69
S & W "100 Years of Gunmaking!" By C. R. Hellstrom	72
S & W Revolvers—Current Prices	84
Appendixes	93
Index	139
Photographs	97-138

Acknowledgment and Apologia

I MEAN TO TOUCH only on such facts in the personal lives of Horace Smith and Daniel Baird Wesson as have to do directly with their work. It is the life of two Americans seen through the windows of their work.

A bibliography of minutiae has a specialized interest and then again it runs up book production costs.

Acknowledgements of indebtedness are due to Carl R. Hellstrom, President of Smith & Wesson; City Library Association of Springfield, Mass.; American Antiquarian Society, and Maryland Historical Society.

To all who have in any way contributed to the glory of Smith & Wesson through the medium of this volume, I tender grateful thanks.

MARTIN RYWELL

DANIEL BAIRD WESSON

DANIEL BAIRD WESSON WAS BORN MAY 18, 1825 in Worcester, Massachusetts, the fourth of ten children of Rufus and Betsey Baird Wesson. There were five sons and five daughters.

John Wesson, who emigrated from England and settled in Salem in 1644, is believed to be the founder of the Wesson family in this country. The early home of the family was in the New Hampshire colony. Records show the marriage of a Samuel Wesson to Martha Haven, the ceremony taking place May 7, 1711. In the history of Fitzwilliam (Cheshire County) New Hampshire, mention is made of Jonathan and Molly Wesson, and of their children, Jonathan, Polly and Josiah, all born between 1784 and 1786.

Rufus, the father of Daniel Baird, was a grandson of Abel Wesson of New Hampshire, and he himself was born in New Hampshire in 1786. In early manhood Rufus was attracted to Massachusetts by the workshops and some kinfolks and settled in Worcester. Here he met and married Betsey Baird, who came of old and prominent families.

Worcester at the time is thusly described:

"This geographical centre of Massachusetts would present no very striking attractions to the lover of natural scenery. But this valley possesses precisely those features which art is capable of rendering extremely fascinating. And there is scarcely to be met with, in this or any other country, a more charming landscape than Worcester presents, from almost any of the moderately—elevated hills that surround it. The high state of agriculture in every part of the valley, and the fine taste and neatness exhibited in all the buildings of this flourishing town, with the great elegance of many edifices and the intermingling of so many and fine shade and fruit trees, spread over the prospect beauty of a high order, on which the eye delights to linger. I have never seen, in a community of equal extent, so few marks of poverty and human degradation, as in this valley: and it is this aspect of comfort and independence among all classes, that enhances greatly the pleasure with which every true American heart contemplates this scene: since it must be considered as exhibiting the happy influence of our free institutions."

Free institutions were meaningful to Worcester. Her participation in the battle of Lexington was described thusly: "Before noon, on the 19th of April, 1775, an express came to town, shouting as he passed through the streets at full speed, 'To arms! to arms!—the war's begun!' His white horse, bloody with spurring, and dripping with sweat, fell exhausted by the church. Another was instantly procured, and the tidings went on. The bell rang out the alarm, cannon were fired, and messengers sent to every part of the town to collect the soldiery. As the news spread, the implements of husbandry were thrown by in the field; and the citizens left their homes, with no longer delay than to seize their arms. In a short time, the minute-men were paraded on the green, under Captain Timothy Bigelow. After fervent prayer by Rev. Mr. Maccarty, they took up their line of march to the scene of conflict."

Two companies of minute and militia men marched from Worcester to Cambridge, on the alarm, April 19, 1775. A Samuel Wesson and a Daniel Baird were among them.

The negroes of Worcester petitioned the committee of correspondence to assist them in obtaining their freedom. It was resolved in a convention held in Worcester, June 14, 1775: "That we abhor the enslaving of any of the human race, and particularly of the negroes in this country, and that whenever there shall be a door opened, or opportunity present for anything to be done towards the emancipation of the negroes, we will use our influence and endeavor that such a thing may be brought about."

It was in the autumn of 1775 that a group of volunteers under the command of Arnold engaged in a winter march through the wilderness on an expedition against Quebec. In that attack on December 31st, a Sergeant Wesson of Worcester was slain.

A Worcester newspaper account of July 24, 1776, describes the scene when the Declaration of Independence was received on Saturday, July 14, 1776: "On Monday last, a number of patriotic gentlemen of this town, animated with a love of their country, and to show their approbation of the measures lately taken by the Grand Council of America, assembled on the green, near the liberty pole, where, after having displayed the colors of the thirteen confederate colonies of America, the bells were set ringing and the drums a beating: After which, the Declaration of Independence of the United States was read to a large and respectable body, among whom were the selectmen and committee of correspondence, assembled on the occasion, who testified their approbation by repeated huzzas, firing of musketry and cannon, bonfires, and other demonstrations of joy; when the arms of that tyrant in Britain, George the III, of exercrable memory, which in former times decorated, but of late disgraced the court house in this town, were committed to the flames and consumed to ashes; after which, a select company of the sons of freedom, repaired to the tavern, lately known by the sign of the king's Arms, which odious signature of despotism was taken down by order of the people, which was cheerfully complied with by the innkeeper, where the following toasts were drank; and the evening spent with joy, on the commencement of the happy era.

1. Prosperity and perpetuity to the United States of America. 2. The president of the Grand Council of America. 3. The Grand Council of America. 4. His excellency General Washington. 5. All the Generals in the American army. 6. Commodore Hopkins. 7. The officers and soldiers of the American army. 8. The officers and seamen in the America nnavy. 9. The patriots of America. 10. Every friend of America. 11. George rejected and liberty protected. 12. Success to the American arms. 13. Sore eyes to all tories, and a chestnut burr for an eye stone. 14. Perpetual itching without the benefit of scratching, to the enemies of America. 15. The Council and Representatives of the State of Massachusetts Bay. 16. The officers and soldiers in the Massachusetts service. 17. The memory of the brave General Warren. 18. The memory of the magnanimous General Montgomery. 19. Speedy redemption to all the officers and soldiers who are now prisoners of war among our enemies. 20. The State of Massachusetts Bay. 21. The town of Boston. 22. The selectmen and committees of correspondence for the town of Worcester. 23. May the enemies of America be laid at her feet. 24. May the freedom and independency of America endure, till the sun grows dim with age, and the earth returns to chaos.

Rufus Wesson found Worcester to his liking. He became a manufacturer of wooden plows and they known for their excellent construction in the carving of the convexed curves to turn the furrows. About 1836, with the advent of the cast-iron plow and the demand for them, Rufus abandoned his wooden plow manufacture and took up farming. Daniel was then eleven years of age.

Until he was eighteen years of age, Daniel resided at home, devoting his time equally between duties on the farm and schooling, slighting neither, yet nursing a hope that he might soon be free to follow the bent of his inclination. His father seemed to think that the shoe business afforded a fine prospect for him and urged him to master it under his brothers, Rufus and Martin. Incidentally, Martin Wesson did so well that he opened a shop as a bootmaker on Main Street in Worcester in 1849.

While Daniel did not like shoemaking, he was constrained to enter it, but he soon found it uncongenial and went back upon the farm. There he essayed some boyish pistol-making, with the old flint-lock of his father as a model. Wooden stocks patiently whittled and barrels molded from abandoned vessels of pewter were deftly put together and fearlessly tested.

Daniel hoped to be sent off to the shop of his brother, Edwin, but his father did not readily entertain the notion of a second departure, and in the end Daniel had to pay for his time to gain his freedom. He was eighteen when he made this bargain and finding that his father valued his time until attaining his majority at one hundred and fifty dollars, he paid him that sum out of his savings and went off to join his brother, Edwin.

Edwin Wesson was the oldest brother. He had served his apprenticeship under Silas Allen, manufacturer of flintlock rifles and pistols in Shrewsbury, Massachusetts. About 1837, Edwin Wesson went into business for himself at Hartford, Connecticut, with Joshua Stevens and S. C. Miller under the firm name of Wesson, Stevens, and Miller. They manufactured percussion revolvers under the patent granted Daniel Leavitt April 29, 1837. The partnership of Wesson, Smith and Miller was dissolved about 1849.

Joshua Stevens had been employed by Samuel Colt. Colt received his U. S. patent February 26, 1836, and organized the Patent Arms Manufacturing Company of Paterson, New Jersey, to manufacture guns under his patent.

In 1849 Edwin went into business with his brother, Daniel, at Northboro, Massachusetts, to manufacture the Leavitt revolver. They also manufactured a percussion sporting rifle. Thomas Warner was associated with the Mass. Arms Company as an employee.

Thomas Warner, in his testimony in the trial of Samuel Colt versus The Massachusetts Arms Company of June 30, 1851, describes himself and his relationship to the Massachusetts Arms Company as follows:

"I first commenced learning the millrights' trade; worked at that a year or two; then I commenced filing locks in the United States Armory at Springfield, in 1810; I filed locks eight or ten years; went from there to operate the water shops; worked at various branches there; I went to other water shops and worked a few years and then came back again; in 1827 or '28, I was appointed inspector of the work at the upper water shop, and of the tools and machinery; in 1833, I was appointed United States inspector of contract arms; served in that capacity until the summer of 1837—July or August—and was then appointed Master Armorer of Springfield Armory; served in that capacity until 1843; then I went to New Haven and took a bid to get up the tools and machinery for Mr. Whitney, for the manufacture of the United States rifle, the so-termed new model percussion gun, in January 1847, or about that time,—it may have been in December 1846—I commenced in January; the bargain was made with Mr. Colt to manufacture one thousand of his revolvers in connection with Mr. Whitney; Mr. Whitney owned the establishment. These revolvers were made at the armory there; they were made under my supervision with instructions from Mr. Colt. In April, 1849, I went to Hartford with E. G. Ripley, Esq., administrator of the estate of Edwin Wesson to finish out a lot of fine rifles (150 or so) which Mr. Wesson had partially commenced. Mr. Wesson was a rifle-maker. I finished that job, and the Massachusetts Arms Company hired me to see to getting up the tools and machinery for the manufacture of their arms; I am still with them; I was hired by them in October or November, 1849. I have no interest in the patent; I was hired by the Company, on wages, just as I was by Mr. Colt."

In answer to a direct question, "Are you a stockholder in this company?" (Mass. Arms Co.) Warner replied, "No, Sir."

Edwin Wesson died in 1849 and a corporation known as the Massachusetts Arms Company was organized March 5, 1850. It succeeded to the business and the patents which included Edwin Wesson's patent of August 28, 1849 and September 22, 1845. A pending application for a patent at the time of his death was granted August 28, 1849 to Edwin J. Ripley of Hartford, administrator of the estate of Edwin Wesson. It was for an improved revolver with a mechanically operated cylinder and the frame pivoted upward forward of the hammer to permit the quick removal of the cylinder for reloading.

The Massachusetts Arms Company continued to manufacture the Wesson-Leavitt revolver until stopped by Samuel Colt in an infringement suit in June, 1851.

In September, 1848 Samuel Colt had written to President Polk that the Leavitt patent is an infringement and has prejudiced some of the officers of ordnance against the Colt revolver.

Colt had obtained a basic patent on February 25, 1836 with sweeping claims. In the specification for the patent he recites: "Among the many advantages in the use of these guns, independent of the number of charges they contain, are, first, the facility in loading them; secondly, the outward security against dampness; thirdly, security of the lock against the smoke of the powder; fourthly, the use of the partitions between the caps, which prevent fire communicating from the exploding cap to the adjoining ones; fifthly, by the hammer's striking the cap at the end of the cylinder no jar is occasioned, deviating from the line of sight; sixthly, the weight and location of the cylinder, which gives steadiness to the hand; seventhly, the great rapidity in the succession of discharges, which is effected merely by drawing back the hammer and pulling the trigger.

The basic claims Colt makes are as follows:

"I claim as new—1. The application of the caps at the end of the cylinder. 2. The application of a partition between the caps. 3. The application of a shield over the caps as a security against moisture and the action of the smoke upon the works of the lock. 4. The principle of the connecting rod between the hammer and the trigger. 5. The application of the shackle to connect the cylinder with the ratchet. 6. The principle of locking and turning the cylinder. 7. The principle of uniting the barrel with the cylinder by means of the arbor running through the plate and the projection under the barrel. 8. The principle of the adopter and the application of the lever, neither of which is used in pistols."

Daniel's association with Warner soon terminated and for the next two years Daniel is hard to follow. He joined his brother, Franklin, who had a gun-making establishment at New England village in the town of Grafton on the southeastern boundary of Worcester.

Franklin Wesson received his training as an apprentice of Samuel T. Hawken, a famed manufacturer of rifles. Frank became a gun manufacturer in his own name. He manufactured sporting and target rifles, a double-barreled rifle, a carbine and a tip-up pistol for which he received a patent October 25, 1859. An interesting oddity of his manufacture was a superposed pistol in which the barrels revolve and a sliding dagger is in between the barrels. It was alleged that criminals carried this superposed dagger pistol in their vest pocket and put lead pencils in the barrels to mislead the police. Gilbert H. Harrington later became associated with Franklin Wesson in Wesson and Harrington which was succeeded by Harrington and Richardson. William A. Richardson had worked for Frank Wesson in 1866.

Daniel left his brother Franklin's employ to become superintendent of the Leonard Pistol Manufacturing Company at Charlestown, Massachusetts. As to the identity of the Leonard Pistol Manufacturing Company, it was owned by George Leonard, Jr. Tradition has it that Daniel Wesson was employed for several months by George Leonard, Jr., to help develop the Leonard pistol. Leonard had been employed by Allen and Thurber in the manufacture of pepperboxes and conceived the idea of a modified action for which he obtained patents, September 18, 1849 and July 9, 1850. It was a hammerless percussion ring trigger pepperbox which was manufactured for Leonard by Robbins and Lawrence. The 1850 patent (U.S. 7493) in the specifications refers to the barrels being "drilled and bored entirely through," but Leonard did not follow this specification through by asserting it in his invention claims. If he had it would have had the validity of the Rollin White patent issued several years later, which gave Smith and Wesson a cartridge revolver monopoly.

It is therefore possible that as a result of Daniel Wesson's employment with Leonard and the subsequent manufacture of the pistol by Robbins and Lawrence that we find the motivation that led Daniel Wesson to employment with Robbins and Lawrence. From Robbins and Lawrence Daniel entered the employ of Allen, Brown and Luther (Frederick Allen, Andrew J. Brown and John Luther) gun barrel manufacturers in Worcester, Massachusetts, where he met his subsequent partner, Horace Smith.

Daniel Wesson then worked for George P. Foster at Taunton, Massachusetts. Foster manufactured Plains rifles made for the Western trade; also the first American bolt-action rifle. This was a bolt-action breech-loading needle rifle manufactured under the Klein patent of November 12, 1849. Foster also manufactured the Porter rifle, which was patented October, 1951, by Patrick W. Porter. It was a percussion nine-shot, pill-lock with a steel cover that fit over a vertical or turret cylinder. Colonel Porter was killed by the back-fire while demonstrating the weapon to Samuel Colt. An experiment about this time came very near costing Daniel Wesson his life. Colonel Porter persuaded Daniel to assist him in a demonstration of the rifle of his invention before a board of ordnance officers. Notwithstanding every precaution in handling it one of the chambers went off independently, sending a bullet whizzing through Daniel's hat; while another chamber, pointed directly at his body, narrowly missed fire.

The 1851 Springfield, Massachusetts Directory lists Daniel Wesson as living on Morris Street and employed by Springfield Arms Company. The Springfield Arms Company is described as "A new establishment. The building is a few rods above Rice's building. It manufactures Warner's Patent Revolver pistols and carbines. They are the invention of James Warner, formerly of South Hadley, who is agent of the company. The guns are six shooters; they have twenty thousand dollars worth of machinery which consists of fifteen engine lathes, planing machines, upright lathes, and all the appurtenances of a pistol factory. The number of hands employed is sixty. Also manufactures carbines. Turns out twenty-six pistols daily."

The same director also lists the following item which should be of interest in view of Smith and Wesson's later development of the rim-fire cartridge.

"Percussion Cap Factory. This concern carried on by James Chattaway and is in Rice's building. The process of manufacture is quite interesting, and the article produced second to none in the world. The copper which forms the cap is purchased from the rolling mills, of suitable thickness for working, and the sheet first passes through a machine which cuts it up into pieces, or 'blanks,' of the form of a cross, the four arms being of equal length. One of these pieces contains the material for a cap. They are then taken up, one by one, by girls and pressed into a die which gives them their complete form. They are then placed in a frame with holes in it for their reception, the cap opening upwards. A frame with corresponding holes, each hole being charged with the mercurial detonating powder, is then placed on the other, and a slight causes the powder to fall through into the caps, and they are primed. The caps, still remaining in the frame, are then taken to another machine, for the purpose of dropping into each, a drop of waterproof cement. This is accomplished by a frame of pencils which descend into the cup of cement, and retaining a drop, each, are shoved directly over the cap frame, and then lowered into the corresponding caps beneath where the cement is deposited. The cap is then finished, and is put up in tin boxes for the market. Mr. Chattaway furnishes about 100,000 caps a day. The boxes are made in Hartford. Six qualities of caps are produced, varying in weight from 4 to 12 ounces per thousand. There are eleven dies in the room, and the number of hands employed is ten. The process of manufacturing is a beautiful one, and considering how every little cap is handled singly,—marvellously rapid."

HORACE SMITH

HORACE SMITH WAS BORN OCTOBER 28, 1808, in Cheshire, Massachusetts, son of Silas and Phoebe Smith, one of eight children. When he was four years old, his father, who was a carpenter by trade, moved with his family to Springfield, Massachusetts, where he found work in the United States Armory. The family lived in a small house near Blake's woods.

When Horace was 16 he went to work as an apprentice at the United States Armory in Springfield. He started as a helper to a bayonet forger in which position he remained for four years, then he served in other departments until after eighteen years' employment Horace became a master gunsmith.

Then he left the Armory and went to Norwich, Connecticut, where he worked for a year with Allen & Thurber. Charles Thurber was a partner with his brother-in-law, Ethan Allen, in the manufacture of firearms. Thurber had received a Master of Arts degree from Brown University in 1827, taught at Milford Academy, and for eight years was principal of the Latin Grammar School in Worcester. It is possible that Daniel Baird Wesson as a student had met Charles Thurber. In 1840 Thurber became a partner in Allen & Thurber.

Ethan Allen was a gunsmith in Bellingham, Massachusetts, when Benjamin and Barton Darling lived there. The Darling brothers secured a patent for the pepperbox April 13, 1836, which claimed the rotation of the cylinder by cocking the hammer as did Samuel Colt in his patent of February 25, 1836. Allen set up a shop in Grafton, Massachusetts, and manufactured pepperboxes under his patent relative to a method of raising the hammer and driving it down with one pressure of the trigger or a double-action mechanism.

Horace Smith came to work for Allen and Thurber in 1842. Thurber, who had mechanical ability, was busy completing his invention of a hand printing machine. It was the first invention that approximated a typewriter in the modern sense and in 1843 he received a patent for his premature invention. This is another example to prove that an inventor cannot take time by the forelock.

Horace Smith left Allen and Thurber and worked for several months at the armory of Eli Whitney at New Haven, Connecticut. Smith then returned to work for Allen and Thurber at Norwich, Connecticut, and remained there until 1846, when he went into business for himself.

Smith set up a gunsmith shop in Norwich of Cranston and Smith for experimental work and gun manufacture. He manufactured a pistol invented by Orville Percival in 1850 and also the 22 caliber Flobert pistol which had been developed in France. Behind the Flobert pistol was an interesting story and an important development in a breech-loading cartridge.

In Europe when dueling was in flower, practice pistol ranges were popular and the 22 caliber pistol the favorite target practice weapon. In such a gallery, about 1840, Flobert, a Parisian pistol instructor, invented a breech-loading metallic cartridge by crimping a round 22 caliber bullet to a percussion cap. The fulminate in the copper cap propelled the bullet but there was no powder charge. The charge of fulminate in the head was the only propellant of the lead ball. About 1850 Flobert had developed his invention into a rim-fire cartridge with a powder charge. This became the famed "bulleted breech cap" or BB cap which developed into the "conical bullet cap" or CB cap, a slightly heavier conical bullet cartridge.

In 1849, when his business was slow, Smith worked for a while for Oliver Allen, in Norwich, who was manufacturing whaling guns.

Courtlandt C. Palmer hired Horace Smith to do some experimental work to simplify the Jennings design of December 25, 1849. The result was U. S. Patent #8317 granted Horace Smith on August 26, 1851, for an "Improvement in Breech-Loading Fire-Arms" which he assigned to Courtlandt C. Palmer. The patent claims were as follows:

"What I claim as my invention and desire to secure by letters patent, is operating the breech-pin directly by the finger-lever, as herein described, in combination with the breech-pin and abutting lever, formed and operating substantially as herein described and for the purpose specified.

"Second, I also claim elevating the charge-lifter by direct contact of the breech-pin carrier with an arm of the lifter-lever and depressing it by the direct contact of the finger-lever with the other arm of the said lifter-lever, as described."

The patent had the tubular magazine under the barrel and the ring trigger of Jennings, but Horace Smith's improvements are concealed. It is in the interior mechanism that we find the differences.

A wedge-shaped breech-block, which narrowed in the direction of the butt, was used in place and instead of the notched rack, whose forward end acted as a magazine gate.

Another refinement was that the muzzle end of the breech-block was depressed by a link action of the ring trigger. It moved backward and the magazine spring pushed the hollow-base bullet backward and then upward. A forward motion of the ring trigger inserted the bullet in the chamber of the weapon and closed and locked the action.

Later Smith discontinued his business and went to work for Allen, Brown and Luther, where he met Daniel Wesson, who was there employed.

Smith and Wesson were to find another employer in common and that was Courtlandt C. Palmer, who financed the experimental work in improving the breech-loader.

III
THE PATENT BACKGROUND

AT WHICH END SHOULD A GUN BE LOADED—at the breech or at the muzzle? It appears to be a simple question, so simple that the reader may think it a foolish one. The answer however represents centuries of work by numerous unsung inventors in unique and ingenious mechanisms to develop the breech-loader. An examination of the products of their futile efforts will excite our esteem and admiration.

Loading a gun at the breech was attempted in the 1300's, almost simultaneously with the introduction of gunpowder in Europe. Numerous attempts followed. In 1664, just to cite an example, we find Abraham Hall has obtained a British patent for a "gun or pistoll, which hath a hole at the upper end of the breech to receive the charge, which hole is opened or stopped by a piece of iron or steel that lies along the side of the piece, and movable by a ready and easy motion." A breech-loader that dates from 1471 is in the British Museum of Artillery at Woolwich. Between 1836 and 1873 over 540 patents were issued in the United States for inventions covering single shot, breech-loading small arms. This figure of 540 does not include pistols nor does it include 120 patents granted during the same period for magazine guns.

Let us turn back the pages of the United States patents to trace the historical continuity of Horace Smith's invention from its immediate ancestry.

The pedigree begins in 1848, August 10th, when Walter Hunt was granted patent #5701 for a "loaded ball" which he titled the "rocket ball." It was a conical lead projectile, the cavity in the base was filled with powder, and the base closed by a disc having a hole in the center to admit the flame from an independent priming unit. This "rocket ball" preceded Colonel Minie's invention and also had the property of expansion into the rifling when fired.

Readers in general of this book may want what Erasmus Darwin, Editor of *The Botanic Garden* requested of James Watt when he wrote: "If you will at leisure hour tell me what the world may know about your improvements of the steam engine, or anything about your experiments, or calculated facts about the power of your engines, or any other ingenious stuff for a note, I shall with pleasure insert it (in *The Botanic Garden*) ... I wish the whole not to exceed 2 or 3 quarter pages, and to consist of such facts, or things, as may be rather agreeable; I mean gentlemanlike facts, not abstruse calculations, only fit for philosophers."

For those interested in detailed technical aspects, Hunt in his patent specification wrote: "It will be readily perceived that this plan of a combined ball and cartridge is well adopted to firearms made to be charged at the breech, from the fact that in firing this ball the cap A is left in the breech, with its head or disk pressed firmly against the breech-plug, and its rim or flange forced outward against the inner periphery of the caliber of the barrel, forming an air-tight stopper, which effectually prevents all backward escape of the powder, notwithstanding the breech-plug may be loosely fitted in the breech of the barrel behind the charge, which must necessarily be the case in all reciprocating breech-pins in order to prevent their binding from heat, deposits, etc., in rapid firing. It is also obvious that in guns made with piston breech-pins, the cap A in my plan would be carried out forward of each succeeding ball, operating as an effectual wiper to the barrel at each discharge of the piece."

In his invention claim Hunt added: "The construction of a ball for firearms, with a cavity to contain the charge of powder for propelling said ball, in which cavity the powder is secured by means of a cap inclosing the back end of the same, substantially as described."

On August 21, 1849, Walter Hunt was granted Patent #6663 for a gun for his "rocket ball." It was a lever-action breech-loading repeating gun with a tubular magazine under the barrel which he designated the "Volition Repeater." This arm was originally patented in England December 10, 1847, and it was in direct lineage with the Winchester 1866. It had a straight drive, spiral-spring-driven firing pin and tubular magazine under the barrel with a complicated repeating mechanism.

Hunt's patent claim for his "combined piston-breech and firing cock repeating gun" was, "1. The construction of a hollow sliding or piston breech-pin, which is operated by a lever in loading and securing the charge in the breech of the gun, which breech-pin, in addition to the above characteristic, contains or has attached to it the mainspring, firing cock or punch, and firing-chamber of the priming. 2. I also claim the plan of transferring the priming from the fixed magazine to the firing-chamber in or by means of the said sliding breech-pin, as above set forth and described."

Walter Hunt was born in Martinsburg, New York, on July 29, 1796. He became a machinist in his home town and remained there until he was thirty when he decided to come to New York City. Here, as an inventor and mechanic he continued to work until his death in 1859. Versatile and prolific, his inventions included a sewing machine, a fountain pen, a flax spinning machine, a heating stove, the safety pin, an iceboat, a nail-making machine and other devices. His lock-stitch needle perfected in 1832, preceded the patent of Elias Howe by a dozen years, but Hunt failed to patent his invention. Hunt fits the stereotyped pattern of the inventor, original and ingenious, but lacking in the ways of practical promotions. That he left to others and therefore Hunt assigned his "rocket ball" and "volitional repeater" to George A. Arrowsmith of New York, a model maker and promoter, who had a small machine shop and in his employ a gunsmith named Lewis Jennings.

Jennings simplified the repeating mechanism, improved the design and was granted U. S. patent #6973 on December 25, 1849, which he assigned to Arrowsmith.

Jennings made the following patent claims: "In combination with a magazine for containing the cartridges or loaded balls, and which communicates with the barrel, the employment of a sliding charger operated substantially as herein described, for the purpose of forcing the cartridges as they are required towards the rear end of the magazine, as described.

"*Second*. I claim making the charger in two parts connected by a spring, and working substantially as herein described, whereby any difficulty arising from irregular working or yielding of the parts will be avoided, and by which, also, the transfer of the cartridges or charges to the carrier is insured.

"*Third.* I claim combining the carrier, the breech-pin, and the abutting or stop-lever, with the sliding trigger-bar, substantially as herein described, whereby all the movements of all these parts are effected by the motions of the trigger-bar, as described.

"*Fourth.* I claim the longitudinal fillet on the trigger-bar in combination with the pinion having one cog grooved for the passage of the said fillet, substantially as described, by means of which the pinion is made to retain the sliding breech-pin in place while the trigger-bar completes its motion to discharge the piece and to elevate the stop or abutting lever, as described.

"*Fifth.* I claim the stop which prevents the passage of the cartridges from the magazine when this is combined with the carrier and magazine, substantially as described.

"*Sixth.* I claim, in combination with the receiving chamber and carrier, the lever which hugs and steadies the cartridge or ball therein, substantially as described.

"*Seventh.* I claim, in combination with the carrier that elevates and transfers the cartridges or charges, the spring catch by means of which the carrier can be held down to permit the piece to be re-cocked without transferring a charge to the barrel, substantially as described. And, finally, I claim the spur on the spindle of the cock in combination with the catch on the sliding breech-pin, substantially as described, by means of which the pull on the cock has the effect to withdraw the breech-pin from the breech of the barrel as described."

The gun used hollow loaded bullets carried in a tubular magazine below the barrel, fed to the barrel by the action of a ratchet operated by a ring trigger which had a forward and backward slide of about three inches, and fired by a Maynard tape primer held in a priming magazine on the top of the frame.

Arrowsmith then sold the patents of Hunt and Jennings to Courtlandt C. Palmer, a New York capitalist, in 1850. Palmer was President of the Stonington and Providence Railroad of Connecticut and a leading hardware merchant in New York City.

Palmer contracted with Robbins and Lawrence of Windsor, Vermont, to manufacture five thousand Jennings patent rifles.

Robbins and Lawrence was one of the larger gun manufacturers and rapidly growing. The firm originally commenced when Richard S. Robbins and Nicanor Kendall formed a partnership in 1843. About a year later S. E. Robbins joined the partnership and the firm was known as Robbins, Kendall, and Lawrence. In 1847, Kendall retired and the firm continued as Robbins and Lawrence.

Palmer subsidized Horace Smith, Daniel Wesson, and Benjamin Tyler Henry in an effort to produce the successful breech-loader. He was a capitalist with faith and vision. He believed that the improvements would lead to the eventual culmination in a practical firearm.

IV

SMITH & WESSON

In 1852 Palmer abandoned production of the Smith-improved Jennings weapon because it was too complicated and lacked power. Palmer financed Smith and Wesson, who continued to work together in an effort to develop a practical magazine arm. It is believed that Benjamin Tyler Henry assisted them.

Palmer had contracted with Robbins and Lawrence of Windsor, Vermont to manufacture five thousand Jennings patent rifles. Daniel Wesson was employed by Robbins and Lawrence and so was Benjamin Tyler Henry.

Henry, a master machinist, was born in Claremont, New Hampshire March 22, 1821. He attended school until he was sixteen and then served his apprenticeship with J. B. & R. B. Ripley, gunsmiths in his town of Claremont. Here he worked on, among varied firearms, a "waterproof rifle." "It was a underhammer magazine patented by R. B. Ripley, Lebbeus Baily and William B. Smith, February 20, 1839, and received U. S. patent no. 1084. The magazine was in the butt. It was a development of Hall's breech-loading rifle.

Henry left Ripley Brother employ and went to work in the Springfield Armory. Here he worked until 1842 when he went to work for Nicanor Kendall & Co. in Windsor, Vermont. Windsor was nearby his home town and Henry's family originally came from Windsor. He also probably knew Nicanor Kendall because Kendall manufactured a stud-lock patented by William B. Smith.

Kendall originated the underhammer percussion hammer. In 1843 when Nicanor Kendall & Co. merged with Robbins, Henry went along with his employer.

In 1850 when C. C. Palmer arranged for Robbins & Lawrence to manufacture the Jennings rifle, it was inevitable that Henry by virtue of the experience in his background should help with the improvement of the mechanism. It is probable that Henry and Wesson became acquainted in this connection. Henry may have gone on to Norwich with Smith & Wesson when they formed the limited partnership.

Out of their combined labors, Smith, Wesson, and possibly an assist from Henry came the invention for which on February 14, 1854 Horace Smith and Daniel Baird Wesson received U. S. patent #10,535.

It was an elevator attachment that fed the bullet-cartridges from the magazine into the chamber. The ring-trigger was replaced by a lever composed of the trigger guard. This lever was pivoted forward of the trigger and the lever operated the loading mechanism. The lever moved back a long cylindrical breechblock which in turn moved back and cocked a center-hammer and the breechblock front became the firing pin.

It was a simple, important and practical mechanism to move and lock the bolt in position with its head to support the cartridge. The claims of the patent were as follows:

"*First.* The combination of the percussion hammer, the sliding breech-piece and the barrel, substantially as described, whereby the breech-pin H, shall serve both as a breech to hold the cartridge in the barrel, and as a means of conveying the force of the blow of the hammer to the cartridge, substantially as set forth.

"*Second.* In combination with an extractor for drawing the shell out of the chamber, a device arranged to strike the shell or cartridge and expel the same from the arm, substantially as described.

"*Third.* The combination of the sliding breech-piece H, and the carrier D, constructed substantially as described, whereby the carrier can be operated while the breech is closed, and also be made to expel the shell while being raised, as set forth.

"*Fourth.* The combination of the hammer and the breech-piece H, with the operating lever G, whereby the hammer may be cocked by the movement of the operating lever, substantially as described.

"*Fifth.* A reciprocating breech-piece arranged to move in line with the bore of the barrel, in combination with an extractor which advances from behind the cartridge to engage therewith, for the purpose of withdrawing the shell from the barrel when the breech is opened.

"*Sixth.* The combination in the breech-loading arm of a hammer, a reciprocating breech-piece, and an extractor with an operating lever, the said parts being arranged to operate substantially as described, whereby the movement of said lever shall open the breech, cock the hammer, and withdraw the shell, as set forth.

"*Seventh.* The combination substantially as described, of a barrel open at its rear end with a reciprocating breech-piece moving in line with the bore of the barrel, and an extractor arranged to advance from behind the cartridge to engage with and extract the shell.

"*Eighth.* In combination with a barrel open at the rear, a device arranged to operate on the shell as it is withdrawn from its chamber, and expel it from the arm, the combination being substantially such as described.

"*Ninth.* The combination substantially as described, of an operating lever, and extractor, and a device arranged to operate on the shell after the breech is opened and expel it from the arm, substantially as set forth.

"*Tenth.* The combination, substantially as described, of an operating lever with a hammer, breech-piece, and extractor, whereby the movement of the lever once forward and back will open and close the breech, cock the gun, and withdraw the cartridge-shell.

"*Eleventh.* The combination, substantially as described, of an operating lever with a hammer, closing breech-piece, extractor, and a device for throwing out the shell, whereby the motion forward and back of the lever will open and close the breech, cock the gun, withdraw the shell, and expel it from the arm."

On June 20, 1854 Smith, Wesson and Palmer formed a limited partnership under the firm name of Smith and Wesson. They owned the Hunt, Jennings, Smith, and Smith and Wesson patents. Smith and Wesson entered into a "Patent Right Deed and Covenant" with Palmer, the essence of which was contained in the following provision: "It shall have the exclusive use and control of all patents and patent rights which the said

Smith and the said Wesson or either of them can or may hereafter obtain or acquire for inventions or improvements in firearms or ammunition or upon the matters already patented as aforesaid, including all power of granting licenses, conveying shares and rights, receiving rents and royalties, and recovering and collecting damages for infringements."

In substance the agreement was that any patent of Smith and Wesson would rebound to the partnership control or benefit. With that the partnership took over the Smith factory in Norwich, Connecticut. They manufactured pistols, and ammunition for them. They also made an experimental rifle in caliber 50.

The Smith & Wesson weapon which was produced at Norwich in 31, 36 and 44 caliber was their patent of February 14th or St. Valentine's Day patent. It used a bullet with a hollow-base and the base contained fulminate of mercury to ignite by detonation and to propel the bullet. It was similar to the system of the needle gun. The Jennings weapon used the Maynard tape primer. In shape, the cartridge was cylinder-conoidal, shaped, with the detonator and propellant in one mixture. The points were cut flat and a pierced cardboard disc was glued to the base of each to reduce the hazard of setting off the adjacent cartridge since each arm held eight to ten charges.

The ammunition was that described by Hunt in his 1848 patent except that the priming mixture was substituted for compressed powder. It was a cylindro-conical bullet weighing about 115 grains with a deep cavity in the back filled with a priming mixture. It was sealed off with a copper washer having a cork disk in the center and was discharged by the impact of the hammer on the breech or firing pin. This projectile with fickle fulminate went off at unpredictable times and it was not effective ammunition because the hollow base could not store enough powder for effective energy.

To load: At a point near the forward end of the magazine tube, the tube was cut through and a sleeve formed around it which contained the magazine spring. The magazine was compressed and held with a catch to insert the cartridge base down. When the tube was filled, the sleeve was turned until it aligned with the rest of the tube and the spring was released.

Smith & Wesson found that fulminate of mercury was a good servant in small calibers but when placed in caliber 44, it became the master. It blasted the chamber walls rather than direct the bullet from the muzzle onto its destination. Small caliber was the answer. Fulminate of mercury also left a very corrosive residue in the barrels. That meant a new barrel since the Smith and Wesson barrels could not be rebored as the old muzzle-loader each time a little larger.

Daniel Wesson tried a new approach with a cartridge similar to the Pottet shotgun cartridge patent issued in France. The case had powder in it with a cup set in the middle of the base. The cup contained a small piece of metal, on top a flake of fulminate and on top of that another piece of metal and powder to almost fill the case but enough space for the bullet. This arrangement was also similar to the metallic cartridge invented by Dr. Edward Maynard, a dental surgeon of Washington, D. C. which ignited by an outside primer. Whereupon Daniel Wesson added beeswax for lubrication between the powder charge and the bullet. (U. S. patent #11,496 granted August 1854.)

In 1854 Smith & Wesson received a Gold Medal for their Patent Repeating Pistol exhibited at the Baltimore Exposition of that year and awarded by the Maryland Institute for the Promotion of Mechanical Arts.

Unfortunately the records of the Institute were all burned in the great Baltimore fire of 1904 but The Book of the Exhibition issued in 1854 contains the following entry:

"A Pistol by Smith & Wesson, deposited by J. Carrier. This specimen is a self loader, and works remarkably well; it is a good invention, and we think it will be extensively used, and is well entitled to notice."

V
VOLCANIC REPEATING ARMS COMPANY

SMITH AND WESSON SOLD OUT in June 1855 to a newly organized Connecticut corporation—the Volcanic Repeating Arms Company. The corporation paid the partners sixty-five thousand dollars in cash plus 2,800 shares of stock for the partnership assets including the plant machinery.

The Volcanic Repeating Arms Company was capitalized at $150,000.00 with 6,000 shares of common stock at $25.00 per share. The stock subscribers were mainly from New Haven, Connecticut. Eighteen out of the 29 subscribers were from New Haven. The subscribers were: James N. Adams, John E. Alcott, Charles B. Bates, George B. Bishop, E. B. Bowditch, C. B. Bryant, R. N. Buchanan, H. T. and C. S. Bushnell, Hiram Camp, James B. Campbell, Paschal Converse, Charles Dickerman, Nelson H. Gaston, Sachett Gilbert, Charles S. Harris, William C. Hicks, Henry Hooker, Green Kendrick, James N. Linsley, Silas R. Perry, J. W. Pose, William H. Robertson, Thaddeus Sherman, Ruben Stannad, I. C. Stock, William Strong, Samuel L. Talcott, Morris Tyler and Oliver Fisher Winchester. Winchester subscribed to 80 shares.

Nelson Gaston of New Haven, one of the largest stockholders, became the president of Volcanic Repeating Arms Company. His background was many years experience in the mining business until in 1854 he had switched to the shipping business. It was late in 1855 when the organization was complete. Temporarily Horace Smith acted as plant manager but took no active part in the new firm. He soon withdrew and moved to Springfield where he kept a livery stable with his brother-in-law, William Collins, on Market Street. Wesson then became superintendent and after a brief period resigned and left. The letter of acknowledgement of Wesson's resignation reads as follows:

New Haven, February 8, 1856

Daniel Baird Wesson,

Dear Sir:

By vote of the Board of Directors of "The Volcanic Repeating Arms Company" I am hereby instructed to inform you of their acceptance of your resignation of the office of Superintendent of said Company, to take effect on Monday next. And also acknowledge their appreciation of your services as a mechanic, and the conscientious discharge of your duties as a man. With respect, I am, Very truly yours, Samuel L. Talcott, Secretary.

Palmer had completely withdrawn, Smith withdrew after a short time, now Wesson left. The reason is difficult to fathom except that a state of turmoil existed for a while because Nelson Gaston, the president, died suddenly in 1856 and Oliver Fisher Winchester became president and the firm had difficulty in becoming firmly established. This is but feeble conjecture.

In February 1856 the Volcanic Repeating Arms Company moved to a small building near the corner of Orange and Grove Streets in New Haven. William C. Hicks, a mechanical engineer was hired as plant manager and about fifty were employed. Employment was under a contract system used by Colt. The most skilled employees were given contracts to assume responsibility for the production of specific items within the plant. A contractor had to be at his bench at seven in the morning just as the workmen. The contractor engaged to fabricate from stock given him, a certain number of a certain part. The contractor hired hands by the day or piece and he was responsible that the work pass inspection as well as for tool breakage and stock wastage. Some of the parts were contracted out to Whitney Arms Company and others.

Gaston, the president, died in 1856 before the firm had become firmly established and Oliver Fisher Winchester became president.

Oliver Fisher Winchester was a small but active stockholder who was impressed by patent possibilities and had some experience as an inventor though he knew nothing about firearms. He had a record of success in business.

Oliver Fisher Winchester and his twin brother, Samuel, were born November 10, 1810 in Boston, the youngest in a family of five children. Oliver's mother was the third wife of his father, Samuel, who had ten children by his two previous marriages. One year after Oliver and his brother were born, the father died. The widow without resources found it difficult to manage.

Oliver, age seven, went to work on a farm and when he was 14 he was apprenticed to a carpenter. He served his time until he was twenty when he became a master carpenter. He then went to Baltimore and for the next three years was a building contractor. He left the contracting business and found employment in a local dry goods store. Aged 24, he married and opened his own retail men's furnishing store in Baltimore. Three years later he opened a larger men's furnishing store and there continued for the next ten years.

In 1847 he applied for and secured U. S. patent No. 5421 for a method of cutting men's dress shirts to prevent pull on the neckband. In 1848 when he secured the patent he moved to New Haven and started a shirt factory. In 1849 he formed a partnership with John M. Davies, a leading jobber of men's furnishings in New York City. They built a larger factory building in New Haven. Eight hundred people were employed at the factory to cut the shirts and the cut shirts were given to about five thousand workers at home to hand sew the shirts.

In 1853 the factory was again enlarged and the sewing machine was introduced. Winchester impressed by the production possibilities of the patent had secured the rights for New Haven County and installed five hundred sewing machines.

Despite this highly successful enterprise, something in firearm possibilities appealed to him and thus in 1856 we find Winchester the head of a new enterprise—the Volcanic Repeating Arms Company.

A line of carbines with 16, 20 and 24 inch barrels were added to the line of the caliber 36 navy pistol with 6 and 8" barrels that were manufactured. The firearms were not successful since they leaked gas at the breech. The answer of the metallic ammunition was still lacking.

The Volcanic Repeating Arms Company lacked working capital since most of the cash went to pay off the Smith and Wesson partnership purchase price. Winchester advanced and guaranteed loans for more than twenty-five thousand dollars but the firm's losses continued. On February 18, 1857 the Volcanic Repeating Arms Company was declared insolvent.

On March 15, 1857 Winchester took over Volcanic Repeating Arms Company for thirty-nine thousand dollars which covered his secured claims. This wiped out all the stockholders including the stock held by Smith, Wesson and Palmer.

Winchester organized the New Haven Arms Company to assume the assets of the Volcanic Repeating Arms Company. It was organized April 3, 1857 under Articles of Association with a capital of fifty thousand dollars divided into two thousand shares of twenty-five dollars par value. Eleven joined with Winchester which included seven shareholders of defunct Volcanic Repeating Arms Company. Winchester was paid forty thousand dollars for the equipment and right to manufacture under the patent assignment which Winchester retained. He received eight hundred shares plus twenty thousand dollars cash or a total of forty thousand dollars.

The reason for the decision of the three men with firearm background, Smith, Wesson, and Palmer to sell out and abandon their patents is as difficult to explain as is the faith of Winchester, a highly successful shirt manufacturer, to persevere in a field he knew nothing about.

VI

SMITH & WESSON AGAIN

SMITH AND WESSON FORMED A SECOND PARTNERSHIP after Wesson left Volcanic Arms Company during their reorganization. Daniel Wesson sought to construct a revolver that would utilize the ammunition he had devised.

The partnership rented a location at No. 5 Market Street between State and Sanford Streets in Springfield. This was probably in 1856 since the 1857 Springfield Directory describes the new firm as "Manufacturers of Superior Revolving Firearms, and Ammunition."

Wesson designed a revolver with a bored-through cylinder to be breech-loaded with a cartridge. Wesson was estopped by a basic patent that had been granted to Rollin White of Hartford, for chambers bored through the cylinder, on April 3, 1855. This bored-through cylinder would permit cartridge insertion in the rear of the chamber. Colt revolver chambers did not extend through the cylinder and the Colt was loaded from the front with loose powder and a separate bullet. Some paper cartridges with bullet attached were used but these were fragile.

Paper cartridges were known as early as the sixteenth century. At first they were merely an individual charge of powder rolled in a paper tube. Later the ball or bullet was attached. It developed into the cylindro-connoidal projectile introduced in 1855 to supplant the spherical ball.

Daniel Wesson at first probably designed the bored-through cylinder because of his familiarity with the George Leonard bored-through cylinder which preceded the White patent. It is possible that when Wesson attempted to ascertain its patent possibilities that he uncovered the Rollin White basic patent.

On November 17, 1856 Smith & Wesson acquired the Rollin White patent with the exclusive right to use a cylinder bored through from front to back. They agreed that as long as Rollin White defend his patent he would receive fifty cents royalty on all guns made by Smith & Wesson. White is alleged to have spent over seventy thousand dollars to fulfill this portion of the agreement which exceeded the amount of royalties he collected.

The Colt patent would expire in the Fall of 1857. Smith & Wesson joined the waiting vigil ready to produce. Remington stood by with the Beal patent revolver, Allen & Wheelock waited with a new revolver but none were so thoroughly prepared for this new competition as Smith & Wesson. They had a revolutionary type arm and had obtained a monopoly for it when they acquired the Rollin White patent which gave them exclusive right to use a cylinder bored from front to back.

The Rollin White patent gave Smith & Wesson a monopoly of cartridge revolvers until 1869 when the patent expired. Congress voted a renewal of the patent but President Grant vetoed the bill because he contended that this patent control had deprived the Union armies of cartridge revolvers during the War Between the States.

Rollin White was born at Williamstown, Vermont, June 6, 1817. He was a very versatile inventor. His patents include the White Sewing machine which is still manufactured; one of the first automobiles, the White Steamer, the name of which still exists as the White Busses and Trucks; a torpedo, cartridges and numerous mechanical devices. His interest in guns was due to his work from 1849 to 1852 as a contractor for Samuel Colt. This contract system provided that the contractor hire employees by either the day or the piece but he must fabricate a certain number of a certain part from the stock given him. The contractor used the facilities of Colt but was responsible that the work pass inspection, responsibility for tool breakage and also stock wastage. It took ingenuity to beat that system.

In Lowell, Massachusetts a firm assumed the name of Rollin White Arms Company without his authorization whereupon he caused them to change their name to Lowell Arms Company. Competition, fair and otherwise arose when the Smith & Wesson revolver proved successful and had a monopoly. Imitators included E. A. Prescott of Worcester, and James Warner of Springfield, Massachusetts. Warner was a former employer of Daniel Wesson, and made a cartridge cylinder with a loading gate. Allen & Wheelock made a cartridge revolver with a side-hammer mechanism. L. W. Pond of Worcester made a tip-up revolver that was hinged immediately in front of the hammer. J. P. Lower of Philadelphia made a 32, some with his name and some marked W. L. Grant or D. D. Cone. Daniel Moore of Brooklyn made a revolver with a swing-out cylinder though later he evolved a ludicrous effort to evade the Smith & Wesson patent control which he called the National revolver and which used a teat cartridge. Colt Patent Firearms Company tried the Theur patent ammunition similar to the teat but it never went beyond the experimental stage. Cartridge pepperboxes were brought out by Elliot, Sharps and Rupertus. Everyone wanted to get in on the act when they saw the performance of the Smith & Wesson revolver and its reception.

The story is told that at a country fair a man sold cider and was kept busy dispensing it. A customer complained that the price was too high. The cider-seller confident of his monopoly replied that the customer could have all the cider free if he could buy it cheaper at the fair grounds. Shortly thereafter the customer returned and said that he had found cider selling at one-half the price. The cider vendor investigated. He found that a person had stationed himself at the back of the tent where his barrel stood, put a gimlet through the canvas and into the barrel head and was filling glasses with cider at half price. All patents have a tapee with a gimlet. Behind the farcical details and astringent humor is the undertones of the tragedy of invention.

Constantly some litigation kept Rollin White stirred up. He died in 1892.

The White action had gas escape at the breech, the ever-present evil which Wesson sought to remove. Smith had manufactured the 22 caliber Flobert pistol which used the BB cap or metallic cartridge and it contained no powder but was propelled by a charge of fulminate in the head. Wesson thought that the BB cap was the answer but the empty shells were difficult to extract after firing because they bulged upon discharge.

The problems that Wesson wrestled with are set forth in the specifications of U. S. Patent No. 27,433 issued to Smith and Wesson for improvements in filling metallic cartridges. It reads:

"We are aware that a metal cartridge for breech-loading pistols ... has been made in which the fulminate is spread in a thin layer over the interior of the base of the cartridge and is held in place by a washer of thin metal or other material.

"The explosion of the cartridge (in our type of revolver) from the hammer causes the base to bulge out—by which the cylinder is jammed and prevented from revolving freely.

"Metallic cartridges have also been constructed with a milled washer inserted in their base and the fulminate contained between the projection and depression around the edge of the washer and the interior surface of the cartridge at its base; but these cartridges are not adapted to the cylinder used in our arm.

"Metallic cartridges have also been constructed with a hollow flanged annular base and the fulminate contained in a hollow ring which is inserted in the hollow annular base . . . but this description of cartridge is expensive and the construction dangerous from the difficulty of closing and turning the ring after the fulminate is introduced without explosion."

Daniel Wesson had tried the Flobert breech caps but the heads of the cartridge bulged and jammed between the cylinder and the frame. The cartridge Wesson developed required "anvils" to explode it. The solution was simple as Wesson developed it. In the cartridge a space where the base joins the column, the shell was filled with fulminating compound to ignite the powder charge. The hammer struck the cartridge edge and pinched it against the cylinder wall and exploded the fulminate. Thus Wesson achieved the rim-fire cartridge. It is simplicity incarnate yet so are many other practical inventions. The obvious is the least apparent. Inventors seek fruition in round about routes and complex circuits and overlook the most likely probable solution.

Smith & Wesson began operations with twenty-five workmen. They had located on Market Street in Springfield to enable them to attract skilled gunsmiths from the Springfield Armory. Smith had spent eighteen years in the Armory and knew the ability of its employees. The erection of a National Armory was authorized at Springfield, Massachusetts by an Act of Congress approved April 2, 1794. President George Washington had selected two sites for national armories, Springfield in the North and Harpers Ferry in the South. An arsenal and powder magazine had already been established at Springfield in 1777 with Washington's approval. It was used during the Revolutionary War for preparation and supply of munitions of war and ordnance of all kinds to the Continental armies. Therefore the establishment with the existing installations as a starting point, the armory was soon completed. The first gun lock hand filed at the Springfield Armory in 1795 by Alexander Crawford took three days since those early arms were entirely handcrafted and the entire lock made by one craftsman. Smith & Wesson began to produce the first revolvers and cartridges in 1857 and applied the principles of the division of labor and interchangeable parts.

Smith & Wesson decided to produce the revolver and ammunition in 22 caliber because it was the safest rim-fire cartridge with the weak copper cases dependent on its own tensile strength at the moment of ignition—three grains of powder and thirty grains of lead. Copper lacked better methods of annealing and production to prevent rupture of a copper case when fired and yet thin enough to be dented by the hammer and set off the detonating compound in the base of the shell.

The revolver was the tip-up model which Smith & Wesson continued to manufacture until 1869. It had a bronze frame and the barrel was jointed to the frame at the front of the top strap. On the bottom it was held by a spring latch arrangement which was released by upward pressure. A circular plate on the left side of the frame was held by a screw which was the axis of the hammer and its assembly. The frame was nickel-plated, the cylinder and barrel were blue and square-end walnut stock was piano-finished. Rifling consisted of three grooves with lands of equal width and pitch of one turn in twenty-four inches, lefthand twist.

Black vulcanite velvet-lined cases with hinged cover were made for the first model, caliber .22, tip-up revolver, with a representation of the revolver on the cover.

In order to bring the chambers of the cylinder in alignment with the barrel, we find in a groove at the top of the upper strap of the frame, a straight spring which used pressure on the slots in the cylinder by means of a lug. Pressure was removed by raising the hammer. The hammer was of two parts and jointed. This cylinder stop method patented by Smith & Wesson in 1859 was complex and unsatisfactory but they were hesitant in using the Colt expired patent of the straight-lever type cylinder stop.

It had an octagon barrel with a rib on its top; a lunar-shaped front sight and a rear sight which was a notch in the uptilted end of the cylinder stop spring. Beneath the barrel was the ejector rod for extracting cartridges. To prevent the action from jamming if a cartridge would bulge at the head, there was a circular recoil shield which contacted the cylinder by means of a lug. The seven-shot cylinder engaged the frame by a knob which was held in place at the rear by a projecting column. An extra cylinder was provided to be loaded in advance and take the place of the empty cylinder. It had a stud trigger.

The revolver failed to have a half-cock or safety notch and Smith & Wesson explained that "a half bent is entirely dispensed with, as it is found more convenient and safe to carry the hammer resting between two of the cartridges; when so placed it is impossible to be accidently discharged."

Therefore we find the cartridge era of ignition ushered in by this small caliber revolver. It was indeed a major achievement though the improvements were not conspicuous in their time. Smith & Wesson gave us the synthesis of these improvements and we have the first cartridge revolver.

The question of priority is misleading. Separated inventors oftimes make similiar independent inventions. The laurel of recognition and the reward we bestow upon the first to translate concept into concrete reality.

This raise the question, what is invention? There has never been a statutory definition of the term "invention," nor indeed has there been "any affirmative judicial definition of that vexatious term." Once a judge referred to an invention as the result of "a flash of genius." At once protest flared up as to who was to define "a flash of genius."

It was Smith and Wesson that ushered in our modern cartridge era of ignition. They were craftsmen who received their training as apprentices and without formal scientific training. Yet when the problems presented themselves it enkindled inborn talents and brought forth the answer.

They were both from New England. The poor soil drove many to enterprise and invention. They turned to firearms, a Yankee industry because Connecticut and Massachusetts had copper and iron.

VII

THE RIM-FIRE METALLIC CARTRIDGE

No ONE PERSON INVENTED THE METALLIC CARTRIDGE just as no one person invented any firearm. Invention is usually a process of gradual development. This development consists of a step by step progress to remove the mistakes that unfold in the process. Unpredictable kinks develop as the human relationship enters into the usage of the invention. Removal of these kinks are the progressive advances that lead to the final perfection of invention.

Smith and Wesson gave us the Rimfire Metallic cartridge. Let us give a fleeting glance over the historical background so that we may appreciate the significance of the contribution.

A firearm is a device which propels a missile. A firearm is a projector of bullets. Bullet comes from the French word, "boulette," a small ball. The bullet is the missile. Propulsion, in turn, is caused by combustion within the firearm. Propulsion of the bullet is a thermodynamic process.

Ammunition is expendable and includes a bullet, the propellant, a primer to set fire to the propellant, and a cartridge case to contain the propellant. This projectile or bullet is also known as the shell or shot. What happens when the trigger or gun control sets the firearm into operation? The propellant is ignited and transformed into a complex gas.

Historically—it was necessary to ignite the propelling charge by touching it off with a burning match. Later developed the MATCHLOCK or a mechanical arrangement for lowering the match into the priming mixture when the trigger is pulled. Then came the use of the flint in the place of the smouldering match and that was the WHEEL-LOCK and FLINTLOCK ignitions. In the 19th century flint was superseded by chemicals and we have the percussion ignition.

The paper cartridge came into use in Europe during the last quarter of the 16th century and remained the cartridge standard until Smith & Wesson. Paper cartridges were made by enclosing the bullets and the powder charge in a roll of paper and tying the ends or sealing them with paste. Even pig's intestines were used as cartridge case containers.

The conical bullet supplanted the round ball when rifling became accepted. Rifling is increasing the accuracy by spinning the projectile about its longitudinal axis. How? By grooving the barrel. The barrel or tube of the gun is so basic an element that the Germans and the French each call the gun bore by their word for "soul." The shallow grooves along the bore are the rifling. Of smaller diameter they are called lands and those of greater diameter are the grooves. The lngth of the bore for one complete turn of rifling is called the twist.

In the beginning of the 19th century we find experiment with conical bullets, because of rifling. The conical cartridge expanded too much and would not spin. Captain Minie in his invention of the bullet used an iron cup in the hollow base which was driven forward by the discharge to insure expansion.

Then they tried to make a bullet the exact shape of the bore. It was still paper cartridges but now using conical bullets instead of round balls. Christian Sharps in 1852 introduced a linen-self-consuming cartridge. Linen held its shape better and was more durable than paper. The case material didn't matter because it was generally impregnated with compounds to make the case itself burn or explode from the flame of the discharge. Samuel Colt collaborated with the Ely Brothers of England to improve his patent on a self-consuming cartridge. It was made of paper impregnated with a mixture of potassium nitrate.

Paper cartridges with a partial ignition system were used in the Franco-Prussian War. The Germans in their needle-gun cartridge and the French in their Chassepot cartridge. The german cartridge used a priming pellet in front of the bullet. The firing pin had to penetrate the entire powder charge to fire the cartridge and was therefore known as the needle. The French chassepot also used a pellet of priming compound but at the cartridge base.

Therefore with rifling came the conical bullet. The next advance in cartridge design came with Percussion ignition. Between 1797 and 1804 French chemists sought a substitute for saltpeter (see 1935—The .357 Magnum Chapter in this book). Out of their experiments came recognition of the explosive qualities of fulminate of mercury and chlorate of potash. These chemicals exploded by percussion or a sharp blow. Out of these observations, Alexander Forsyth, a Scotch Clergyman, developed the percussion ignition system in 1807.

The percussion ignition gave rise to putting the detonating powder in small copper tubes. The tube or cap became the basis of the metallic cartridge primer. Several metallic cartridges were designed but none with their own ignition system.

Dr. Edward Maynard, a dental surgeon of Washington, D. C. who invented the tape primer magazine, claimed his metallic cartridge offered the solution for the breech-loader.

Pauley, a Paris gunsmith, in 1812 had patented a cartridge with a priming base and a charge of powder contained in a metal cylinder that was screwed onto a bullet. (Brevet d'Invention 843, old series). The powder charge and projectile were contained in a combustible envelope of paper, but this envelope in turn was seated in a brass "rosette." This resembles the brass head of a modern shotgun cartridge. A pellet detonating compound was placed in a primer pocket in the head of this rosette to form a center-fire, self-igniting cartridge with a combustible paper case, but also with a brass head which could be reloaded by inserting new loads and priming pellets. Napoleon encouraged Pauley. It was Napoleon who once said, "Le feu est tout; le reste un peu de chose" or Fire-power is everything; all else is trivial. "In a more philosophical mood Napoleon had observed," Inventions that are before their age remain useless until the stock of general knowledge comes up to their level. "Muzzle-loading weapons had to give way to an increase in breech-loading. Then came the quest for the metallic cartridge because the cartridge case is the only breech seal. The copper or brass case expands under the pressures developed through the ignition of the powder charge. This expansion causes the case to cling to the walls of the chamber and seals it against gas escape to the rear. The brass case of Pauley's cartridge was thick and would not expand to seal the bore as the powder gases expanded and the primers did not fit their sockets gas-tight.

It was the same situation as Samuel Colt with the repeating mechanism had to coincide with percussion. A rifle that fired twenty-four times in succession without reloading, by a circular movement of its magazine, was approved by the Academy of Sciences of France in 1767. It was the invention of a father and son Bouillet at Saint-Etienne, in Forez. Tests showed that after eighteen rounds were fired in a minute and a half, the barrel became almost too hot to hold, and it was found advisable to wait two minutes before firing the remaining shots. The committee thought the gun could be fired better if the shooter wore a glove. Milot was introduced before the French National Assembly on April 14, 1792 as the inventor of a machine of war capable of firing fifty shots a minute. In August 1792 a French mechanic Renard invented a gun that was capable of firing ninety times in a second. A German named Gass and he was the chief chemist at the porcelain factory at Serves claimed the invention of a pistol which would fire six to fifteen bullets a second. There were others before Samuel Colt but the time was not ripe. The time must be propitious.

The time must be propitious. Pepys' Diary has an entry of July 3, 1662, which we find the idea of the repeating gun, was known.

"Dined with the officers of the Ordnance, where Sir W. Compton, Mr. O'Neile, and other great persons were. After dinner, was brought to Sir W. Compton a gun to discharge seven times; the best of all devices that ever I saw, and very serviceable, and not a bauble; for it is much approved of, and many thereof made."

And again we find another entry in Pepys' Diary of March 4, 1664.

"There were several people trying a new fashion gun brought by Lord Petersborough this morning, to shoot off often, one after another, without trouble or danger."

The time must be propitious.

After Pauley, who gave us the first type of self-igniting cartridge, came the pin-fire cartridge of LeFaucheux of Paris (1836) perfected by Houllier. A number of French patents followed Pauley's invention, including one by Pottet. It was a tapered-wall center-fire cartridge with a gas-tight primer, but it was handmade and prohibitive in cost since brass drawing by machinery was still unknown. The most popular was the pin-fire cartridge.

Houllier in 1847 perfected a pin-fire cartridge which exploded the primer with a pin inserted at the side of the cartridge base. It contained within itself the percussion cap, the powder, and the charge of shot, with a small brass pin impinging on the percussion powder and attached to it. When the pin was struck, it discharged the gun. Houllier also patented a rim-fire cartridge. Lang, an Englishman, invented a breech-loading shotgun with a crank-operated extractor to use the Houllier ammunition. Frank Forrester, an English sportsman, in "The Complete Manual for Young Sportsman," sounded off with a surge of sadistic sarcasm on the question of whether metallic cartridges were practical. He wrote:

"This gun has . . . no possibility of being loaded or fired except with the identical cartridge prepared for it, which is, and can be, only prepared at the shop which supplies the gun. It is true that the cartridge cases remain in the gun, and on the withdrawal can be recapped and recharged many times, but, apart from the inconvenience of lugging about on your person a hundred or two, if you expect a good day's sport, of these cartridges—since the idea of a sportsman sitting down in the middle of a snipe-bog or a cane-brake, to recharge the cartridges out of powder-and-shot magazine, which he must carry about with him, is preposterous—what on earth is the shooter to do, if he takes it into his head to visit the Himalayas, or the Rocky Mountains, Canada, or the Cape, or any other distant shooting ground (by no means impossible to, or unattempted by British sportsmen), where cartridges for Lang's breech-loading double-barrels certainly are not to be found growing on thorn bushes? Is he to carry with him, in heaven's name, a hundred barrels of cartridges on camel-back, or mule-back, or on his own back, with the consciousness that these indispensables, once used up, his double-barrel is of less use, even, than a broomstick?

"The want of simplicity is enough to ruin any invention, and this, it needs no prophet to foretell, must be inoperative, except as a pretty plaything to be used at home."

Forrester concluded that the answer was a muzzle-loader that loaded quickly with a powder flask.

In October, 1899, the *Literary Digest* made the following statement concerning the future of a new invention: "The horseless carriage will never come into as common use as the bicycle." In 1900 there were 4000 automobiles in the United States and by 1950 over 40,000,000.

Some people lacked imagination and faith. Smith and Wesson translated imagination and faith into actual achievement by an incredible amount of hard work.

In England, Hanson and Golden obtained a patent on a hollow-base bullet in 1841 to use fulminate of mercury not only as a detonator, but also as a bullet propellant. Stephen Taylor obtained a patent in 1847 for a hollow-base bullet. The base of the projectile was covered with a cap provided with a touch-hole for ignition of the charge. Taylor explained the requirements for a weapon that could use the ammunition. "The projectiles are to be contained in a magazine consisting of a self-acting feeding tube, which is attached to the barrel of the piece, and as one projectile is discharged, another is to be pulled forward into the breech of the barrel and fired in succession.

Those were the problems in France and England, while in the United States the background began with the Hunt patent.

Daniel Baird Wesson perfected the rim-fire cartridge though in 1854 he patented a center-fire cartridge. Wesson had solved the problem and produced the rim-fire metallic cartridge by January 1858. A cartridge with the priming mixture in the rim of the head and a powder charge as a propellant was Wesson's answer and it became a vital achievement in the progress of firearms.

The rim-fire cartridge had a projecting flange around its base with an annular recess in which the fulminating powder was placed. The fulminate from the central portion of the head was removed. Wesson invented a loading tool that spun the fulminate into the recess in the flange of the cartridge head. The fulminate in place, a wad was inserted into the head of the cartridge and the case filled with gunpowder. An elongated conical formed ball that was grooved at the rear was then pressed into the mouth of the case. Light pressure was then applied to the head to bring the metal into close contact with the fulminate.

Smith and Wesson were fine gunsmiths and their approach was as a famed inventor once summed up, "First get a clear notion of what you desire to accomplish and then in all probability you will succeed in doing it."

The breech-loading loaded waited upon the development of the key. The key was the metallic cartridge with its own primer. The gas-tight joint was no longer a requirement.

In 1704 the French Academy gave its approval to a breech-loading musket. A lever below the cartridge chamber could be pulled down and the ball and powder inserted and then the lever was pushed back into place. Maurice de Saxe, the famous French marshal, invented a breech-loading rifled carbine. Many of these firearm inventions originated in France because 18th century France was a great military power with one of the largest and most efficient armies of Europe.

Rim-fire cartridge was developed from the breech-cap used in the indoor short-range pistols by Flobert of Paris. The BB cap used a small round bullet that was contained in a metal case. A charge of priming compound was placed in the upset rim of the case but no powder. The explosion of the priming compound by the compression of the rim of the case against the breech of the barrel when it was struck by the hammer expelled enough gases to drive out the bullet.

Horace Smith and Daniel Baird Wesson lengthened the case and placed a charge of powder between the primer and the bullet and the rim-fire cartridge was born. It was a .22 short. And out of that cartridge came the modern system of cartridge ignition. They combined the cartridge case, the conical bullet, the detonating cap and the powder. Patents were mere incentives to encourage more invention on how the original patent could be circumvented.

General James G. Harbord of the United States Army once noted that "the roads you travel so briskly lead out of dim antiquity, and you study the past chiefly because of its bearing on the living present and its promise for the future."

Therefore we found by a hasty review that cartridges date back almost to the beginning of firearms. Either metallic or paper, they were only rigid containers of convenience. The introduction of the metallic cartridge was for safety in handling or rapidity in loading or for waterproofing but failed the function for fire leakage and residue clogging. It was these factors of fire leakage and residue clogging that retarded the breech loader development. As for convenience we had the common paper cartridge that was either torn or bitten off the end, the powder poured into the barrel followed by the ball and the paper used as a wadding. We awaited the thin, springy metal container to seal the bore when expanded by powder gases. This involved the form and construction of the projectile as factor number one. Factor two was the relationship between the amount of the propellant and the weight of the bullet. The priming mixture was the third factor. The fourth factor was the strength of the cartridge case. Thus we see that weapons required separate loading of the bullet, propellant and primer. Then came the invention of the copper cartridge case and the internal primer and the gradual development of gas-tight mechanisms which made possible the production of the modern weapon.

We recognize Smith and Wesson as the fathers of the rim-fire metallic cartridge. The difference between the rim-fire and center-fire cartridge is simply in the position of the priming compound in the metallic case. Rim-fire cartridge contains its priming compound around the up-set rim of the cartridge and fires the cartridge by impact of the hammer or firing pin which crushes the rim of the cartridge against the breech of the barrel. The center-fire cartridge contains priming compound in one of a number of the types of primers which are inserted in the center of the case when the cartridge is made. This primer consists of a primer cup in which the priming compound is inserted moist and then covered with a protective coat and inserted in the cartridge. This is so inserted as to bring it in contact with an anvil. The firing pin strikes the CENTER of the primer, the priming compound is crushed against the anvil and the cartridge is fired.

Smith & Wesson manufactured the ammunition and the first caliber produced was the .22 short. A firearm is described by the diameter of its bore which corresponds to the size of the projectile. This diameter is used as a measuring unit called a caliber. Caliber .50 designates a gun having a half-inch bore. The derivation of the word caliber is uncertain. It may come from the Latin, qua libra—how many pounds or what size. It may be derived from the Arabic, "qualib," which means a mold.

In addition to the metallic cartridge, Smith & Wesson had the Rollin White patent of 1855, the basic patent. Much litigation followed but the basic patent was sustained and it is ironic that it was sustained by the United States Supreme Court February 8, 1869 when it expired. (See cases of White vs. Boker, Dist. Court, New York, 1862 and reported in 29 Federal Cas. 981 and White versus Allen District Court Massachusetts 863 and reported in 29 Federal Cas. 969.)

White's application for extension was denied by the Commissioner of Patents on April 2, 1869, but a bill passed Congress which would have allowed a rehearing. This bill was vetoed by President U. S. Grant on January 11, 1870. Another bill was favorably reported by the House Committee of Patents on February 28, 1873, but this bill failed of enactment.

The Army as usual was slow to accept cartridge ignition, but clung to percussion. On December 21, 1866, Captain Fetterman was sent out from Fort Phil Kearney in Wyoming to relieve a wood train. He chased a few mounted Indians into an ambush of several thousand Cheyennes and Sioux. His command of eighty men was split, surrounded and wiped out in forty minutes with the Indians armed only with bows and arrows.

The Fetterman massacre and the Wagon Box fight of August 2, 1867 and several other incidents began to provoke criticism of the Army's continued use of percussion. Officers of Calvary regiments stationed in Indian country began to complain that their arms do not conform with the frontier service demands.

It is not only a long and rugged path in the development of an invention, but a similar experience awaits the invention before we have complete acceptance. The wheels of progress grind slowly, sometimes the recoil throws them backward but they move onward. It is a dual reaction somewhat like the recoil in a gun which is produced by two separate causes. One, it is due to the reaction caused by the forward motion of the projectile, and two, it is due to the reaction of the expanding powder gases after the bullet has emerged.

VIII

1861-1874

1861—32
1865—32
1869—44 American
1870—Russian
1874—5 Schofield

So MUCH LED TO THE WAR BETWEEN THE STATES and so much followed from it that it is the crucial event in our history. For the number of men involved, the amount of space traversed, the coast line blockaded, of material consumed and results achieved, it exceeded all wars up to that time. The North had 23 states with a population of 22 millions. The South had eleven states with 9 million population, of which 3½ millions were slaves. There were about 2½ million men in both armies. More than 600,000 lost their lives, or more Americans were killed in it than in the American forces in both World War 1 and 2 combined.

If, and that always opens wide the portals of fancy for the historian, Smith & Wesson had produced a .44 caliber cartridge revolver, the course of the war would have terminated sooner. The War Between the States was a percussion ignition war. Military changes move slowly as they stumble through the maze of red tape. Smith & Wesson had licensed the manufacture of the cartridge for the Spencer Rifle, patented by Christopher M. Spencer of Connecticut. About 60 million cartridges for the Spencer Rifle was used during the war. The Henry Repeating Rifle used during the war was a further development of the Hunt-Smith patents. The Confederates did not exactly like the Henry Rifle and called it, "that damned Yankee rifle that can be loaded on Sunday and fired all week." The Henry was manufactured by the New Haven Arms Company. During the war the Remington was converted to cartridge and utilized the .46 rim-fire cartridge presumably with Smith and Wesson permission. Charles D. Leet of Springfield, Massachusetts began the manufacture of cartridges under Smith & Wesson patent of April 16, 1860. Union Metallic Cartridge Company, founded in 1867 by Hartley, Schuyler and Graham, manufactured under license from Smith & Wesson.

With the outbreak of the War Between the States in 1861, Smith & Wesson developed a .32 rim-fire cartridge that had more power than the .22, The revolver frame was steel; the cylinder was 6 shot; it did not have a recoil shield and the ratchet formed part of the cylinder; the jointed hammer was eliminated and a flange on the top raised the cylinder stop. These were some of the changes as Smith & Wesson sought practical improvements. A defect was the tolerance between the cylinder and the top strap so that when powder fouling gummed the action, the cylinder had to be turned by hand.

Though not adopted by the government it immediately became popular as a personal side-arm with the Union forces. This new revolver was never sold to the army for issue to the troops but individual officers bought and paid for them. They recognized the superiority of the metallic cartridge. The metallic cartridge made for a fire power far in excess of that in the percussion revolver. Military requirements of the war speeded development and use of the metallic cartridge—all components in one unit.

Smith & Wesson also produced a second model of their .22 and this also became a popular arm in the army. Smith & Wesson could not keep up with their orders. They were growing and the cartridge revolver was a phenomenal success from the start. They had outgrown their rented place on Market Street and in 1860 had moved into a plant that they had erected on Stockbridge Street in Springfield.

The War Between the States abolished a world as we were given over to fratricidal carnage. The subsidence of that great convulsion had left a different country from the one it found, and we may say that the war marks an era in the history of the development of the American nation. It introduced into the national consciousness a new sense of industrial development, of a world being a more complicated place than it had hithertofore seemed with success more difficult. In this rapid changeover and shifting of scenes, science and invention gave great impetus. Big business was entering the stage as small firms existed or combined.

1865—32 1½ Old

The War Between the States after four bitter years came to an end at Appomattox Court House on April 9, 1865, to be followed by a bitter "reconstruction period."

In May, 1865, Smith & Wesson brought out Model 1½. It was improved and now chambered for the .32 short rim-fire instead of the long and for the cylinder stop on the top strap, one inside the bottom strap was substituted.

In May, 1865, Smith & Wesson brought out the Model 1½. It was a rim-fire, six-shot, single action, using .32 caliber long Smith & Wesson cartridge. It had a six-inch octagonal, ribbed barrel rifled with five grooves and its total length was ten and three quarters inches. It weighed one pound, eight ounces. The cylinder was one and three-sixteenths inches long. A white brass front sight was set into the barrel rib; the rear sight, a small V-shaped groove was cut into the top of the cylinder stop. It had a top-break system, hinged in rear of barrel over the top of the cylinder. The cylinder stop was in the frame extension strap over the cylinder. Removal of the cylinder pin permitted the use of the pin for the extraction of shells.

This model was produced until 1868. The .32 was popular for years. "Wild Bill" Hickok had one on his hip on the August day in 1876 when Jack McCall shot him in the back. Hickok was playing poker in a Deadwood, South Dakota saloon and never had a chance to draw the gun. McCall's plea was self-defense and he was released by the "miner's court" of Deadwood. But he was later tried in the United States District Court at Yankton, convicted by a jury and hanged. The poker hand that Hickok held when he died was a pair of aces and a pair of eights. It has been referred to in the West ever since as "dead man's hand."

Amidst this rapidly growing business Daniel Wesson found time to serve in the Common Council of Springfield in 1866.

In February, 1868, Smith & Wesson announced a third model of the .22 which was only a change in eye appeal. One year later they brought out Model 1½ for both .32 short and .32 long. It was the last rim-fire other than the .22 caliber that Smith & Wesson would manufacture. It was introduced in January, 1869 and manufactured until 1875.

Let us pause for a moment to find out what happened to the New Haven Arms Company. It continued to manufacture Volcanic rifles and pistols, added a .30 caliber pistol but these firearms lacked a satisfactory gas seal at the breech. B. Tyler Henry was plant superintendent and he wrestled with the problem of the cartridge. He produced some flanged metallic rim-fire cartridges in the fall of 1858 which he apparently copied from Wesson's cartridge except Henry's was larger. Henry then adopted the bolt and firing pin to load, fire, and extract the rim-fire cartridge. The result was U.S. patent 3446 granted October 16, 1860 and the Henry rifle was a reality. New Haven Arms Company then abandoned Volcanic arms manufacture and began the manufacture of the Henry rifle. The base of the shells were stamped "H" for Henry and this practice has continued to the present time. In 1861 New Haven Arms retooled for this new weapon—the Henry Rifle—but the constant loss and the costly changeover left the company bankrupt. In 1866 New Haven Arms Company became Winchester Repeating Arms Company. Tradition has it that a verbal understanding between Smith & Wesson and Winchester was that Smith & Wesson would not produce rifles and Winchester would not manufacture handguns.

Illustrated London News of August 31, 1867 reported the Paris International Exhibition and commented:

"Messrs. Colt exhibit a case of revolving pistols and rifles. These arms possess no special novelty, and indeed, are behind the age, insofar as there are none of them adapted for cartridges containing their own ignition; while on the Continent capping revolvers are as much out of date as muzzle-loading guns. Messrs. Colt have in some instances, fitted movable stocks to their pistols. If they would go a step further and adapt their pistols for self-igniting cartridges, these stocked revolvers might serve as useful arms for yeomanry regiments and various mounted corps.

"Messrs. Smith and Wesson exhibit revolving pistols and cane rifles, adapted for pistol cartridges. They are the owners of an American patent (1855) which has seriously hampered the application of self-igniting cartridges to revolvers. By this patent Messrs. Smith and Wesson hold the right in America of boring the revolving chambers from end to end.

"Although it may perhaps not be impossible to contrive a pistol adapted to self-igniting cartridges, the chambers of which are not bored from end to end, such an arrangement is necessarily a roundabout way of achieving what Messrs. Smith and Wesson, by virtue of their patent, accomplish much more simply. Thus makers of revolving pistols, desirous to go with the age and to abolish caps, must either manufacture under a license from Messrs. Smith and Wesson or must contrive some system which will render them independent of this patent, and which will yet be as cheap and efficient. To this cause we believe is to be traced the backwardness of Messrs. Colt in this matter. They are unwilling, as we understand, to purchase a license from Messrs. Smith and Wesson; and their efforts to accomplish the application of self-igniting cartridges to revolvers, without infringement of the patent referred to, have not yet met with the success which they are sanguine will ultimately crown them. In countries where Messrs. Smith and Wesson's patent does not hold the system of capping revolvers, as we have noticed, has entirely disappeared."

1869

The Rollin White patent expired in 1869 and the monopoly of Smith & Wesson of cartridge revolver manufacture was at an end. White's application for extension was denied by the Commissioner of Patents on April 2, 1869, but a bill passed Congress which would have allowed a rehearing. This was vetoed by President U. S. Grant on January 11, 1870, on the opinion of the Chief of Ordnance that an extension would compel the government "to pay to the parties already well paid a large royalty for altering its revolvers to use metallic cartridges." Another bill was favorably reported by the House Committee of Patents on February 28, 1873, but failed of enactment.

General William Buel Franklin became general agent and Vice-President of Colt's and served until 1888. He was an able engineer, a trustworthy executive, an honor student at West Point. His military background and contacts helped him with government contracts. At West Point he was a classmate of President Ulysses S. Grant.

General Franklin tried to buy the patent from White in 1866 and the price was apparently too high. Rollin White was entitled to 25c royalty for each pistol manufactured. It would mean that Colt would have to buy both the White contract and the Smith & Wesson contract. The figure was somewhere around $800,000.

Colt's had considered many plans and had made many experiments to circumvent the Rollin White patent. The text of a patent must be closely studied since it may be novel only as to a minor detail and does not cover protection for the creation of an entire object though the patent drawing may illustrate an entire object. Rollin White had a basic patent and not one for an improvement.

Samuel Colt died January 14, 1862. Elisha K. Root, friend and factory superintendent became President of Colt's Patent Fire Arms Manufacturing Co. Root died in 1865. Then Richard W. H. Jarvis, became President. He had a legal background and joined the Colt organization when his sister married Samuel Colt. The active management however devolved upon General Franklin, who joined Colt's in 1865.

Colt had examined the system that Joseph Rider had developed in 1865 for converting the Model 1860 Army Colt, but this system did not find favor with Colt. There was N. C. Locke's safety notch attachment. There was Captain Farley's plan. And there was F. Alexander Thuer, who came directly from Prussia to Colt's factory in 1849 and became one of Colt's most valued employees.

Thuer tried to develop a metallic cartridge. His initial patent was dated September 15, 1868 (U.S. 82,258) and an additional patent January 4, 1870 (U.S. 98,529). Thuer's conversion method and cartridges aided Colts in getting into the metallic cartridge field.

It was a novel mechanism—a free turning of the conversion's cylinder ring. The essential feature of the cartridge cylinder was a plate or ring which fitted and turned independently on a tube with a ratchet on its end formed by cutting down the cylinder back of the stops. The Thuer ring contained a rebounding firing pin and also a lever for ejecting the cartridges loaded by means of the regular loading lever. A turn of the ring with a thumb piece brought under the hammer either the firing pin or the ejecting lever.

Thuer's system infringed on a patent granted to Stephen W. Wood September 20, 1864 (U.S. 44,363) because Wood's patent was the first to cover ejecting shells by hammer. In December, 1867, Colt's bought Wood's patent.

Yet Thuer conversion of Colts did not make their public appearance until 1869 when the White patent expired and even then it was with hesitancy because of the possibility of the extension of the White patent.

1869—American Model
Model #3-1

Smith & Wesson had not built their trust upon the political promises of the extension of the White patent but instead placed their faith in careful preparations of a technological support. They purchased the W. C. Dodge patents nos. 45,912 and 45,983 for a stronger latch to fasten the top strap securely to the frame and improved hinge. They also purchased the Charles A. King patent no. 94,003 for simultaneous extraction of all shells by a rack set flush with the rear of the cylinder. Ejection of empty cartridge shells had been a problem for the metallic cartridge.

The result of the acquisition of the patents was Smith & Wesson's first martial, caliber .44, American model of 1869, a top-break revolver. It preceded Colts by four years in the use of a large caliber revolver chambered for metal cartridges. The Peacemaker .45 came out in 1873.

The Army needed a cartridge revolver of .45 caliber to do away with "the great inconvenience of loading and capping an army pistol while in motion on horseback, and also the great number of pistol cartridges that are destroyed in the cartridge box, and in loading."

The Model #3 was a six-shot, single action with an eight-inch round, ribbed barrel, rifled with five grooves. The top break system was used, operated by a latch in front of the hammer. The barrel was jointed to the frame at the forward end of the bottom strap. The ejector was operated by a rack and pinion mechanism, incorporated in the barrel joint. Breaking the revolver caused the ejector, which was set on a square shaft, to rise in rear of the cylinder and throw out all the empty shells in one operation. The rifling was right-hand twist, one turn in 20 inches, which has remained as the standard for large bore Smith & Wesson arms. The total length was 13½ inches. It weighed 2 pounds, 10½ ounces. The fluted cylinder was 1 and 7/16 inches long. It had a German silver blade front sight; the rear sight is on the latch. The barrel, frame, cylinder and trigger was blued, the hammer case-hardened and the walnut stocks oil finished.

It was chambered for the Smith & Wesson American center-fire cartridge with a 218 grain, outside lubricated, round nose bullet and a 28 grain black powder charge. The cartridge was developed on the Berdan system centerfire with outside primer and a brass case. General Hiram Berdan had invented the .42—a bottle-necked cartridge with an outside primer. General Berdan had commanded a Federal regiment of sharpshooters during the War Between the States. In 1865 he invented and patented a breechblock that altered muzzle-loading percussion rifles to breech-loading and also the ammunition for it. The actual caliber of this model was .429.

In 1870 the Small Arms Board of the War Department met in St. Louis and examined several .44 caliber Remington revolvers, both percussion and altered to cartridge; several National Arms and Whitney revolvers, the Remington .50 caliber single-shot pistol and Smith and Wesson's .44 caliber revolver.

They found, "The Smith & Wesson is decidedly superior to any other revolver submitted. It should be modified as follows, viz: made center fire; the cylinder lengthened so as to close the space in front of the breech-block, and counter sunk to cover the rim of the cartridge; caliber increased to the standard."

The Chief of Ordnance, Brevet Major General A. B. Dyer, recommended the purchase of one thousand Smith & Wesson revolvers. It was a slightly modified version of the Smith & Wesson American model. They were ordered by the Army December 29, 1870 and were delivered in 1871.

The 1871 report of the Chief of Ordnance stated:

"A small number of revolvers (pistols) which use the primed metallic cartridge, have been made and issued to the troops, and the few reports upon them which have been received at this Bureau show that they are greatly superior to the revolvers which use the paper cartridges, and must supersede them in the service."

1870—Russian Model

Grand Duke Alexis of Russia visited the United States in 1869 and was taken on a buffalo hunting trip by "Buffalo Bill" Cody. Herds of buffalo roamed the West. Some herds covered fifty square miles and contained about half million buffalo. Some herds were said to total 12 million buffalo. It was the food, raiment and shelter for the pioneer. "Buffalo Bill" Cody was armed with a Smith and Wesson American Model .44 revolver. Grand Duke Alexis was filled with admiration for the qualities of this firearm and upon completion of the buffalo hunting trip, he proceeded to Springfield to see Smith and Wesson and ordered 200,000 of this model.

The Russians wanted some changes but they concerned the design details of the exterior. They wanted an extrusion on the back of the frame to keep the shooter's hand from sliding upward above the grip. That was minor but they did have some important ideas about the ammunition. They wanted the cylinder to be reamed to twice the thickness of the cartridge wall or 1/100 of an inch. The cylinder should not be chambered to exact diameter of bore and with this change the shell did not need a shoulder. The open end of the shell was crimped around a groove on the bullet. The lead was 264 grains and the powder 23 grains. The bullet had a uniform diameter. This increased the velocity and the lead content but reduced the powder requirement.

The Russian contract was completed by 1875, or at the rate of 175 revolvers per day. Russia went to war with Turkey in 1877 and Russia won. The war over we find the Turks in 1879 buying Smith & Wesson revolvers. The Turkish model was similar to the American model except that it was adapted for a .44 rim-fire cartridge somewhat like the Henry rifle cartridge.

In 1872 Smith & Wesson submitted to the United States Ordnance Department for tests a .45 caliber, six-shot, single-action. It was accepted and six thousand were manufactured from 1875 and supplied to the Army.

1875—The Schofield Model

It was known as the Schofield Model because the revolver embodied the patents of Major George W. Schofield, a United States Army Officer, for auto-ejection.

The revolver was similar in appearance to the Model 1869 American. The top-break system operated by a latch on the receiver, the latch also forming the rear sight. The ejector was set on a rounded shaft and rose in the rear of the cylinder, when the revolver was broken. Later models were improved by a milled top-latch with a hollow rear to facilitate the placing of the thumb for cocking.

The .45 Smith & Wesson cartridge was 230 grain bullet and 28 grains of black powder. It could be used in the Colt .45 revolver but the Colt .45 cartridge was too long for the Smith & Wesson.

Major George W. Schofield, an officer of the Tenth Cavalry, had taken out a patent dated June 29, 1871, on a barrel-latch mounted on the frame rather than on the barrel and designed to be closed by the hammer before firing. He also held a patent of April 22, 1873, on an ejector whose spring was housed inside its stem and on a rotating crank enclosed by the recoil plate.

His brother, Major General John M. Schofield, had been Secretary of War in 1868. George W. Schofield was an ardent student of firearms and as a result of overwork on an invention took his life with his own revolver at Fort Apache, Arizona, on December 17, 1882.

In 1874 Major Schofield presented his modification of the Smith & Wesson Russian model with the cylinder stop arranged so as to allow ready removal of the cylinder; his patented form of ejector was modified to reduce the number of parts; the comb of the hammer made more concave; the safety lip ground off, and the handle knuckle reduced.

The Ordnance Notes Report of 1873 indicated that "As the reports . . . plainly show the superiority of the Colt revolver over all others tried." It was the result of the .45 caliber Colt Army model of 1873 known as the Peacemaker as against the American model Smith and Wesson. The War Department adopted the Peacemaker.

However, after presentation and testing of the Schofield model we find the Board's conclusions of April 22, 1874.

"Resolved, that Major Schofield's alteration of Smith and Wesson's revolver, in consequence of its simplicity, efficiency, strength, lightness, and the ease with which it can be taken apart; and of the probability that its peculiar construction will diminish the cost of its manufacture below that of the Smith & Wesson pistol, is well suited for use in the military service."

The series of tests continued with a trial of loading on horseback by an expert horseman at a hard gallop. The horseman was able to eject six empty shells and reload the Schofield with cartridges taken from a belt pouch in 26 seconds, the Colt required 60 seconds. The Board reviewed its findings on June 30, 1874, as follows:

"1. That the shapes of the handle and of the comb of the hammer of the Schofield pistol are better than those of the Smith & Wesson;

"2. That the parts of the pistol pertaining to the ejector, cylinder-stay, and barrel-latch seem to be strong and simple, and can be easily dismounted, and replaced, and so far are considered as improvements over the Smith & Wesson pistol;

"3. It is thought that the barrel-latch of the Schofield pistol is safer than that of the Smith & Wesson, in consequence of its always being closed by the hammer before the piece is fired."

The weight of the pistol (2 lbs., 8¾ oz.) places it between the Colt's (2 lbs., 5¾ oz.) and Smith and Wesson, old model (2 lbs., 10½ oz.). A portion of this difference is due to the length of barrel, which in these pistols is respectively 7", 7.5" and 8".

In consideration of the foregoing facts, and of the ease with which the pistol can be loaded on horseback at high speed, it was Resolved, "That in the opinion of the Board, Major Schofield's revolver is well suited for the military service, and that the Board do recommend that a limited number of these pistols be placed in the hands of troops for comparative trial with the Colt's and Smith & Wesson revolvers now in service, and that, as far as possible, the different pistols be tried side by side in the same commands."

The Army purchased 3,000 Schofield model revolvers chambered for .45 center-fire with 28 grains black powder, bullet 230 grains. A later model had several changes in that the trigger had less curve, a checkered and altered barrel-latch, and a longer-bearing surface base-pin.

The story had not ended. In 1876 an Ordnance test was arranged with the Schofield Model, Remington .44 or the 1875 model, and the Peacemaker Colt model to arrive at a decision "in view of the fact that no comparative trial or record has as yet been made between the Colt or Schofield." A series of tests were devised. The Schofield took only 59 seconds to fire 18 shots commencing and ending with chambers empty while the Colt and the Remington took one minute, 54 seconds. The results of the other tests and comparisons are given in Appendix #1.

The Board concluded a "decided preference" for the Colt as "best adapted to meet all the requirements of the military service." This report was approved by the Secretary of War in 1876.

These findings and somersault tactics with the change of Board members and administrations produced partisan feeling about the comparative merits of the Peacemaker versus the Schofield. The report was challenged as "so manifestly one-sided and unjust in its conclusions." The Colt's Bridgeport ammunition showing penetration of "pine butts" was challenged as to meaning. The Schofield jointed frame was praised because "by its peculiar construction all the empty shells are ejected at once, leaving the chambers exposed ready for loading without going through the tedious operation of punching them out one by one, as is the case with the Colt."

Advocates of the Colt replied with the praise for the solid frame Colt because there are "fewer parts, no hinge to rust or wear loose, no catch or catch spring to get out of order—a solid frame."

The International Centennial Exposition held at Philadelphia in 1876 awarded the Gold Medal to Smith & Wesson for its military model and noted:

"Commended for its capacity as a military arm. The ejection of all the discharged shells simultaneously, the workmanship and materials used in its manufacture, its interchangeability, and the simplicity of its action, render it a desirable weapon for defense or offense."

In 1887 twenty-five Schofield revolvers were fitted by the Ordnance Department with a safety stop invented by Colonel John C. Kelton, Assistant Adjutant General. It was a thumb-operated safety pin which, unless it was disengaged, stopped movement of the hammer at full cock.

IX

1873—SMITH RETIRED

In July, 1873, Horace Smith sold his interest to Daniel Baird Wesson and retired from the partnership. Daniel Wesson remained and assumed the firm name of Smith & Wesson, which had engendered international goodwill.

An account at the time mentions that Smith "had made a large fortune, and his health was somewhat impaired, so that, with a wise judgment which, unhappily, very few businessmen are able to exert at the proper time, he determined to give rest to both body and mind."

"A great perseverence, a brilliant business success, and now an honorable retirement, are the three distinctive features presented in the life of Horace Smith. Working out his own career with a patient energy which has had few equals, he became not only successful, but famous as the manufacturer of the Smith & Wesson firearms. At present living in retirement from all business in his magnificent residence on one of the hills of Springfield, commanding a wide and enchanting view of Connectict and its scenery, he may well be regarded as a person worthy of affording a striking example among self-made men.

"His ability as a mechanic and business man have permanently left their impression upon the manufacturing interests of New England. Never a bold but always an aspiring man, he produced good results wherever he applied his energies and enterprise. For instance, in the case of his pistol a positive public prejudice against it had to be overcome, and it required penetration of mind in regard to the future, as well as high mechanical knowledge of the merits of the weapon itself. Satisfied on these points, and being ever ready and willing with all the personal application which such an enterprise could possibly require, he embarked in it without a doubt of the success which he finally achieved in connection with his partner.

"Looking back on what he has accomplished, Mr. Smith can see many dark and discouraging hours, but not one in which he ever lost faith in the ultimate success of steady perseverance and unswerving integrity. These were the strong anchors from which he would never let go, and he hopes by his example to induce others to abide by the same resolution."

Smith had served on the Springfield Board of Alderman to which he was elected in 1859 and again in 1862 and 1863.

Upon his retirement Smith devoted himself to the hobby of astronomy, a subject on which he was an enthusiast. He built an exceedingly fine telescope in his home and followed the movements of the heavens. Smith also traveled in the West and invested largely in western lands and projects, being especially interested in Michigan pine land. He had been recognized as a good businessman and was connected with the Chicopee Bank, the Springfield Institution for Savings, and the Worthy and Riverside paper companies. In 1887 he was elected President of the Chicopee Bank.

Smith was a Methodist and through his generosity the Asbury Methodist Church of Springfield was built. In private Smith was both generous and unobtrusive. Many a poor boy owed his education to his benevolence, and many old people were able to pass their old age in comfort through his gifts.

Horace Smith had been married three times; to Miss Eliza Foster, who died in 1836, to Mrs. Eliza Jepson, who died in 1872 and to Miss Lucretia Hebard of Norwich, Connecticut, who died in 1887. His only children were two sons by his first wife, one of whom died in infancy, and the other, Dexter, a well-known manufacturer, died just six weeks before his father. Horace Smith died January 14, 1893, at his home on Crescent Hill, Springfield, where he had lived since 1860. He left no direct descendants.

His will provided that after several bequests to relatives and institutions, the residue of his estate should be used for public purposes, in the discretion of his executors. The executors caused the formation of the Horace Smith Fund, which is still active. The purpose clause of the fund is, "to help young persons in acquiring an education as may be worthy of aid and encouragement and especially and not exclusively and in imitation of the late Horace Smith."

X

1875-1880

1876—Model 38-2-1
1880—Model 44—S. A. New Model
1880—Model 32—Double Action 1
1880—320 Repeating Rifle
1877—Model 38 Double Action 1

AFTER 1875 SMITH & WESSON REVOLVERS were of such diversified nature and embodying so many modifications and improvements that space does not permit of a complete record.

1876—Model 2-1

Colt began to bring out his "New Line" pocket revolvers in late 1873. They were made in .32, .38 and .41 and it was a solid frame with a screw-in barrel. William Mason was Colt's leading inventive genius at the time and it was he that later developed Colt's first double-action revolver. Mason's patent of 1874 used on the "New Line" revolvers had six original features but its two important claims were the construction of the bolt by which the cylinder was locked in position at the rear and a loading gate. This method of locking the cylinder from the rear and its merits are debatable.

William Mason was a pattern maker who was employed by Colt in 1861. One year later he went to work for Remington and remained there until 1866, when he returned to Colts. Here he was employed for sixteen years and was responsible for many important revolver improvements. In 1882 Mason went to work for Winchester.

Since the .38 Colt sold the best, Smith & Wesson entered the field of the .38. Wesson, using his .44 American-Russian cartridge and its successful history as the nucleus, designed the Smith & Wesson .38. The shell tapered slightly from base to mouth, crimped over bullet and omitted shoulder where gases could be wasted. It weighed 150 grains and had a powder charge of 14 grains.

Wesson then proceeded to produce the first .38 model which also was influenced by the Russian Military Model. It had a standard 3½" barrel to fit the pocket, but was available in 4, 5" barrels and the rear sight was a V-notch cut into lug which was part of the barrel catch. The extractor gear was in the barrel joint engaging round tooth rack screwed to extractor stem held against the action of the barrel in opening arm by extractor gear catch. The hammer was a straight side thumb piece. The cylinder was five chambers, grooved, with inserted gas ring. It was counter bored to receive flanged head extractor with ratchet. The center hole for the extractor stem was made square.

This model was manufactured until 1880 and then continued after changes in the ejector. The improved extractor permitted the cylinder to be taken out with a straight pull. The jointed edges of the frame were beveled. The extractor gear catch cut was eliminated.

In 1891 the model was again changed in the following respects: The barrel lengths were now also 6" in addition to the 3¼, 4, 5. The trigger was pivoted in frame having slender bowed finger piece. The hammer was a flanged and checked thumbpiece. It was slotted to receive rebounding main spring stirrup and rebounding hammer into the rebound notch. This acted as a safety against accidental discharge. It provided three cock notches, rebound half cock and full cock.

The third notch on the hammer was akin to the N. C. Locke system. Colt's single-action Army revolver of 1872 had the third or safety notch as a feature.

About 7,000 of this model was made for Russian trade and was called the Mexican model. You read it straight the first time. That was to conceal the ultimate user of the model. It had several changes in that the half-cock notch was omitted from the hammer; the trigger was a straight-checked finger piece and the trigger spring was made in a lighter weight to reduce the pull.

1880—44 S. A. New Model

In 1879 Smith & Wesson introduced their new model .44 single action. It was made in 4, 5, 6, 6½, 7½, and 8-inch barrels with both long and short straps for the 1 7/16 and 1 9/16 cylinders, six-shot chambered. The extractor catch was first modeled as gear with spring-jointed catch and an engaging round rack. This was changed to the hook type with spring-jointed catch engaging extractor post. This in turn was changed to the hook type with sliding cam catch engaging extractor post. This model was chambered for .44 Smith & Wesson Russian, .44 Smith & Wesson Russian Gallery and .450 Webley. Detachable extension stocks were available.

A question that had not been resolved was whether a revolver could equal the shooting accuracy of a pistol. Ira Anson Paine, about 1880, had defeated all competitors and had established a record as an expert pistol shot with a Lord Model single-shot Stevens pistol.

Paine then went abroad to challenge the masters of the Old World. Here he met Gastinne Renette of Paris. Renette was a well-known maker of target and dueling pistols who ran a gallery in Paris where the French aristocracy practicised pistol shooting against the time when they might have to fight a duel. Renette suggested to Paine the Smith & Wesson .44 single-action with the Russian cartridge as a revolver that could shoot like a pistol. Paine put on shooting exhibitions with the .44 Smith & Wesson and was decorated by the King of Portugal for his astounding skill.

The Standard American Target was ten ring which measures 3.39 inches at fifty yards and Paine's record was 841 out of a possible 1000 in 100 shots. From Boston came W. W. Bennett and his brother Fred, to challenge Paine's record and Fred established a record of 877.

Paine was not happy to recline on his laurels but sought ways to regain his title. He collaborated with Smith & Wesson on a new cartridge, the 39/44 for use in the Smith & Wesson .38 revolver and in 1886 Paine broke Bennett's record with a score of 878 achieved with the Smith & Wesson .38. On the successive days of the competition Paine used his .44 Russian and ran his score up to 904. W. W. Bennett shot a 914.

The race became keener and closer as Paine and Fred Bennett met for six successive days to shoot 100 cartridges a day and to use Smith & Wesson revolvers with 6½ barrels and 3-pound trigger pulls. In 1887 the match proceeded until the fifth day when Paine withdrew due to a protest he filed. Fred Bennett became the United States champion revolver shot with 5093 out of 6000. The Smith & Wesson was proven to be target accurate.

The 38/44 developed for Ira Paine differed from the present 38/44. It had 11 grains of powder, 83 of lead, in a long cylindrical case, the nose of the bullet being crimped inside the mouth of the shell. It was a black powdered cartridge, lubricated on the outside but it had enough power to establish an all-time record. Smith & Wesson manufactured 1413 revolvers of the type made for Ira Paine to accommodate the 38/44 with the bore to approximate the .38 Special.

1880—.320 Repeating Rifle

In 1880 Smith & Wesson brought out the repeating rifle. It was more of a conversion of a revolver than a rifle. It was first manufactured in September, 1880 and production continued until 1887. Nine hundred and seventy seven were manufactured.

Its overall length was 25½ inches of which 20 were the barrel. A few were manufactured with 16 and 18" barrels. It used a detachable stock which clamped by means of a thumb screw beneath the butt. It was chambered for a special Frank Wesson rifle cartridge of .32 caliber with the long shell running full length of the cylinder. The bullet was seated inside the shell and was of the conical variety, the nose flush with the mouth of the brass case and around this was poured a special lubricant. It was more powerful than the .32 with 17 grains of powder and 100 grains of lead. It was more of a copy of what Frank Wesson, Daniel's brother, had been putting out and called a "pocket rifle" which was in reality a revolver with an elongated barrel.

Three years later Colt followed with the Burgess-Colt repeating rifle, a lever-action, chambered for 44/30 center-fire Winchester and in outward appearance just like the Winchester Model .73. At the same time Winchester Repeating Arms Company had several experimental revolver models made and sent them to Hartford gunsmiths for opinions. Colt's Patent Firearms Co. stopped Burgess rifle sale and manufacture in November, 1884. This may add up to a subtle way to achieve a meeting of the minds, or was it unrelated occurrences that happened simultaneously.

1877—Model 38 Double Action 1
1880—Model 32 Double Action 1

Smith & Wesson did not bring out their .38 double-action model until 1877. Colt produced one for the metallic cartridge in 1877. The trend to double-action system forced manufacturers to change. In Europe and England self-cocking or double-action guns were almost exclusively used. The double-action principle is old, it permits rapid firing without disturbing the aim but is liable to accidental discharge.

During the War Between the States, Adams & Tranter double-action revolvers as well as French and Belgium double-action revolvers were imported.

Ethan Allen made a double-action, single-shot pistol under a very old patent. Manhattan and other pepperbox manufacturers had a long production record of double-action guns. Cooper, Remington in the Rider model, Starr, Pettingill, and others produced double-action repeating percussion pistols.

Colt, during his lifetime, objected to double-action guns because he maintained that they were more apt to have mechanical trouble, wasted ammunition and were not accurate. In 1876 William Mason developed Colt's double-action revolver and in 1877 they brought out their first double-action revolver.

In February, 1880, Smith & Wesson brought out their double-action revolver which was the .38 and used the cartridge used in the single-action. The trigger was a flanged finger piece held in a normal position by a V-shaped trigger spring operating on the rocker stop. The front sear was joined to the trigger. It was held in that position with a hand pivot, center distance sufficient to give required double-action throw to the hammer. A flat spring that was inserted in the hand acted as a combination hand and front sear spring. This double-action model was well received and with several changes continued to be manufactured for over thirty years.

In September, 1880, a .32 double-action was introduced. It was similar in outline to the .38 except for the irregular plate which strengthened the frame.

XI

1881-1906

1881—44 D. A. 1
1885—44 Winchester Single Action Frontier
1885—38 Winchester Single Action New Model
1887—32/44—44 Single Action New Model
1886—44 Winchester—44 Double Action Frontier
1887—38/44—Model 3—Single Action
1887—38, Safety, New Departure, Hammerless
1888—32 Safety 1
1891—38—3rd Model 1891
1893—22—22 S. S. 1
1896—32-32 1 Hand Ejector
1899—32/20 Hand Ejector 1
1899—38 Hand Ejector—Military & Police
1902—22 M. Hand Ejector 1
1903—32 1 Hand Ejector 1903
1905—22 Single Shot 2nd Model

In 1881 Smith & Wesson brought out the .44 double-action. Though it was called a Navy revolver in the early Smith & Wesson catalog, yet it was not officially adopted by the service. A similar arm was also brought out and called "New Model Army No. 3" and it was not adopted by the military service. It is similar to the Navy model except that it was single-action, had an oval trigger guard and was made in 6, 6½" barrel lengths only.

The .44 Double-Action Model was similar in appearance to the American and Schofield models. The top-break system operated by a latch in the receiver. The ejector was set on a hexagonal shaft and rose in the rear of the cylinder when the revolver was broken. The revolver used the .44 Smith & Wesson Russian cartridge.

It was not well received. Its weight was reduced, a rebated cylinder substituted and it was catalogued as "Wesson Favorite." But it failed to become the public's favorite.

In 1883 Daniel Wesson admitted his son Walter H. and in 1887 his son Joseph to the Smith & Wesson business.

44 S. A. Frontier

In 1885, Smith & Wesson marketed their Frontier Model. Similar to the .44 Russian, the cylinder was ⅛ inch longer, the top strap and receiver lengthened to take the 44/40 Winchester rifle cartridge. Only 2000 were produced and most of them were rechambered for the Russian cartridge.

They were evidently made to compete with the Colt Frontier, but never achieved popularity. Smith & Wesson supplied extension stock for this model.

44 S. A. New Model or Model 3

Seventy-five were manufactured chambered for .38 Winchester Rifle, with long strap and 6½ inch barrel. A model chambered for 32/44 Smith & Wesson Gallery and .32 Smith & Wesson Target was manufactured with 6½ inch barrel until 1910. About 2920 of this model were produced.

44 D. A. Frontier

Designed as .44 D.A. First Model, it was chambered for the .44 Winchester Rifle cartridge with a powder load of 40 grains and a bullet of 200. It was made on the long strap frame with 1 9/16 inch cylinder.

Model 3—38/44 S. A.

This was the Ira Paine model previously described and it was manufactured from 1887 to 1910. It was chambered for the 38/44 Smith & Wesson Gallery and the 38/44 Smith & Wesson Target.

1887—38 Safety—New Departure
1888—32 Safety—New Departure

In 1887 Smith & Wesson introduced the .38 Safety known as the New Departure, the Safety Hammerless and the "Lemon Squeezer." It was designed in .38 caliber and was followed one year later by the .32 hammerless.

The safety feature as developed by Wesson was clever. It was a safety lever which projected through the back strap of the tang. It connected with the safety latch and was so designed that until the pressure was used on the safety lever, the hammer could not be brought back.

C. S. Pettingill of New Haven, Connecticut patented (U.S. Patent 15,388) a hammerless percussion revolver in 1856. It was manufactured in caliber .44. The hammer was concealed within the solid frame. Pressure on the trigger cocked the hammer, revolved the cylinder and fired the weapon, but it did not prove serviceable, as the mechanism was complicated and delicate.

In 1887 the Navy Ordnance Bureau adopted the Colt double-action with the swing cylinder and ordered 5000. A test was then conducted with both the Colt and the Smith and Wesson .38 double-action safety hammerless. The Board concluded: "That the Smith & Wesson revolver has passed all the tests satisfactorily except the rust test, in which it was totally disabled. The Colt revolver has passed all the tests satisfactorily except the dust test, ... The Board is of the opinion that the issue of a limited number of each of these arms would be advantageous as affording a comparison between the double-acting system and the single-action system in use in the service."

When Colonel Elmer Otis of the 8th Cavalry inquired in 1890 whether Smith & Wesson double-action revolvers could be obtained for the service, the Chief of Ordnance stated:

"The amount of funds available is not sufficient to permit filling requisitions for Smith & Wesson double-action revolvers. It is also deemed unwise to issue an untried arm in large quantities before a preliminary field test has demonstrated its serviceability. This department has already made arrangements to issue 100 each of the .38 caliber double-action Colt's new navy revolvers with improved ejector and a .38 caliber double-action safety hammerless Smith and Wesson revolver with 250 rounds of ammunition per revolver for competitive field tests. These arms will be ready for issue in about three months."

The Board later decided in favor of the Colt and then the Chief of Ordnance recommended the Colt .38.

Colt watched the innovation of the "Safety Hammerless." Patents were taken out in the name of C. J. Ehbets, Colt's patent attorney and assigned to Colt. The patents covered a system of concealed hammer revolvers and were granted on Oct. 14, 1884 and April 30, 1889.

Colt's swing-out cylinder production in 1889 was an attempt to match Smith & Wesson's simultaneous ejection feature. It had its genesis in the patents of William Mason in 1881 (Nov. 15th, Dec. 6) followed by his Aug. 29, 1882 patent. Then came Warnant's patent of July 8, 1884, one patent of C. J. Ehbets and another granted to Horace Lord.

1891—38—Third Model

In February, 1891, Smith & Wesson began the manufacture of this model. The first model was manufactured in 1876, but this model was similar to the 2nd model with several slight changes. Three cock notches were now provided—the rebound, half cock and full cock.

1893—22 Single Shot 1

In 1893, Smith & Wesson brought out the First Model single-shot .22 caliber target pistol. It was an interchangeable combination revolver and single-shot target pistol. The frame was the .38 single-action and it was a top-break pistol with the barrel and cylinder removed and a single-shot barrel hinged to the frame. The hammer was a flanged thumb piece with a solid nose. These pistols were numbered with the 38 single-action revolvers and were manufactured from 1893 to 1905. It became a target gun favorite.

1896—321 Hand Ejector

Chambered for .32 Smith & Wesson and .32 Smith & Wesson long, it was made in 3¼, 4¼, and 6-inch barrel lengths and with six-chambered cylinder. It had a side swing yoke jointed to frame forward of the guard and kept in by a yoke screw at the stud swinging out to the left and carrying cylinder on arm.

As a result of Smith & Wesson's patent of 1895 for a modified system of hand ejector, the model evolved. It was the swing cylinder that had its genesis in the Moore revolver and later developed by Colt's patent of 1887, which was limited to the latch to permit the split frame to be swung apart for ejecting and loading.

The cylinder had six chambers, grooved and inserted gas ring. This was counterbored to receive the flanged head extractor with a ratchet to allow the hand to pass by the tooth in carrying. It had circle cuts to receive the end of the extractor, the arms acting as dowels. For the extractor stem, the center hole was made with a solid key. A flat boss raised from the body undersize to close the space between the barrel and the extractor rod.

Smith & Wesson did not follow up this system for other calibers until 1899.

1899—32 Winchester Hand Ejector

This was a twin of the .38 Hand Ejector except that it was chambered for .32 Winchester. It was made in 4", 5, 6, 6½" barrel lengths. The bolt was incased in a slot in the frame with stud extending through side carrying thumb piece. It pivoted at head end engaging center pin located in extractor unlocking cylinder from frame at bolster. This mechanism when in forward position, rear lug of bolt blocked hammer against cocking and with cylinder unlocked also blocking action of bolt made this safeguard a double protection against accidental discharge.

A 1902 model added the forward yoke locking device. A third change in 1909 altered the hammer at the foot, forward of the rebound seat, thereby engaging the notch in the trigger for double-action. In 1915 a leaf spring was placed in the side plate with the head sticking out between the hammer face and the frame when the hammer rested at rebound, as a hammer block. And the only way to withdraw the block was through the full rearward action of the trigger when firing.

1899—1st Model Military and Police 38—also known as the Army, Navy Revolver Model 1899

When the battleship Maine was blown up in Havana harbor on February 15, 1898, it precipitated the Spanish-American War with the use of armed forces in Cuba and Puerto Rico. The revolver did not play an important role but awoke our military overseers to the needs of the service. Smith & Wesson received an order for 2,000 revolvers for the Navy and 1,000 for the Army and Wesson proceeded to develop a gun.

The contract for these guns had several exact specifications though others were discretionary. It was to have a hand ejector, chambered for .38 Long Colt, and left-hand twist rifling.

The Army and Navy Ordnance Boards preferred the .38 weapon but the actual battle experience of the Spanish-American War was that the .38 bullet lacked sufficient force and stopping power to be effective against human targets. The Army returned therefore to the .45 caliber.

Wesson came up with a patent granted October 4, 1898 for an improved design. It had a 6½" barrel, though a 4" barrel model was sold to the trade. It was the first 6-shot, .38 made by Smith & Wesson. It was a double-action and the barrel was rifled with five grooves. The left swing cylinder was operated by a roughened catch set in the left side of the frame band of the cylinder.

Daniel Wesson also proceeded to design a .38 cartridge and came up with the .38 Special which had a powder weight of 21 grains and a bullet with 158 grains.

Smith & Wesson manufactured this hand-ejector model until 1902 when several changes were made. The important one was an improved method of holding the cylinder in alignment with the bore in accordance with a Smith & Wesson patent of December 17, 1901. It meant a forward cylinder pin lock which made the cylinder more sturdy. A round lug was placed on the lower end of the barrel, through which operated a rear-actuated pin.

1902—Model M Hand Ejector

A miniature pocket revolver on the hand-ejector principle was the first Smith & Wesson .22 caliber multishot produced in 23 years. It had a side swing yoke frame. The barrel was taped, round, made in 2¼, 3, 3½", and threaded to screw to the frame. It has a 7-chambered grooved cylinder with a solid gas ring. The hammer was a flanged finger piece with spring insert to operate stop. It had the forward cylinder locking device.

Chambered originally for .22 long rifle, it did not work because when one was fired, the others loosened and clogged the cylinder rotation so a cartridge was developed which was akin to the .22 long rifle, except it was slightly crimped and was known as .22 Smith & Wesson Long. A change in the model came in 1906 with the forward lock was operated by a knob at the forward end of the extractor rod and available in 3 and 3½-inch barrels. Another change in 1906 was an attempt to convert it into a target model with a barrel of 6 inches. Changes were made in the cylinder stop, but it never clicked with the public.

1903—Model 1 Hand Ejector .32
Regulation Police Model

This model was introduced in January, 1903. It was made in 3½, 4¼, and 6-inch barrel lengths, a tapered, round barrel. The cylinder was 6 chambers, grooved with inserted gas ring, counter-bored to receive extractor with ratchet designed to pass tooth in carrying up, circle cut to receive ends of extractor arms, acting as dowels. The hammer had straight sides, checked thumb piece, round pointed swivel hammer nose. A rebound catch jointed to the hammer and operating between hammer and frame through action of trigger spring.

After serial number 19,425, the cylinder stop notches were lengthened to fit the elongated stud slot for better alignment. There were many changes in this model. In 1917 this model was catalogued as "Regulation Police Model." It is still being manufactured (1953) and a 1952 catalogue describes it as follows: Caliber 32 S.&W. Long, 6 shot, 2, 3¼, and 4 inch barrel, sights fixed, 1/10 inch service type front; square notch rear, stocks checked walnut with S.&W. monograms.

"The Regulation Police Model was brought out to meet the requirements of police departments for a light, small, and absolutely dependable revolver. There was an almost equal demand for this gun chambered for the popular 32 S.&W. Long cartridges and for the powerful S.&W. cartridge. . . . The limitation of barrel length to 4" (in the .38) makes the gun exactly right for a holster or pocket sidearm, and either cartridge is sufficiently powerful to answer the needs of self-protection."

1905—22 Single Shot 2nd Model

The combination single shot target and cylinder arm Model 1 of 1893 was discontinued and this model only single shot made to replace it. The hand and cylinder slots were eliminated but otherwise it was the same model minus the cylinder attachment given with the First Model. It was chambered for .22 long rifle.

In 1909 the model was changed to the "Perfected Target Pistol." The hammer was adjusted for rim-fire cartridge; the barrel catch cam cut was made in strap without the inserted shim; the frame was modeled from the .38 double-action perfected revolver and its lock parts were interchangeable with that model.

XII

DANIEL BAIRD WESSON DIED

DANIEL BAIRD WESSON DIED AUGUST 4, 1906, at 5:15 o'clock, in the afternoon. He had been sitting up "and seemed such a spectacle of vigorous age that the yielding to the depression of the day, and the flickering out of life, was wholly unlooked for." The death of his wife on July 18th had affected him. He had been married to Cynthia M. Hawes, daughter of Luther Hawes of Northboro, Massachusetts, for over 59 years. Daniel Wesson was 81 and had spent over a half century in the manufacture of firearms.

Three children survived him, Sarah Jeanette Wesson, who married Dr. George J. Bull of Montreal; Walter Herbert and Joseph Hawes Wesson. A fourth child, a son, Frank Luther Wesson, died before father.

Daniel Wesson found time to be actively identified with various other business interests, being a director in the First National Bank of Springfield, Massachusetts, and the Bigelow-Cheney Wire Company, and was an officer in the water companies of Leadville, Colorado, and Independence, Iowa.

He contributed to various civic enterprises and took a prominent part in local affairs. He was one of the organizers of the Hampden County Agricultural Society. With James Kirkland and Joshua D. Sackett he purchased the estate out of which Edgewood Park was made. He gave to the city of Springfield the beautiful Wesson Fountain designed by Gilbert and Thompson.

He was one of the founders of the First National Bank of Springfield and for many years one of its directors. In political views, he was strongly Republican.

His son, Joseph Hawes Wesson, took over the reins at Smith & Wesson.

The quality of Daniel Baird Wesson and Horace Smith still stamps its impress though men must pass on but their work lives. In South America, the phrase "Smith & Wesson" entered the Spanish vocabulary as a generic term for all pistols.

Smith & Wesson belong to the category of Fitch, Baldwin, Whitney, Colt, McCormick, Edison, the Wright Brothers—and others—men that made things with their hands, men who used imagination and ingenuity, men intuitive and hard-working. Men who found by trial and error rather than by armchair cogitation.

Inventions are a glorious page in the story of America and laid the foundations for our remarkable advance in industry and commerce. A nation's power or wealth is not wholly based upon its inventions but it is a vital factor.

What factors made for invention? There was the increase of education. Many received their technical education in their training as apprentices. As education increased, invention followed. Encouragement by the government in granting patents which allowed a period for the economic exploitation was another factor. Interest in invention was tied in with the interest in the betterment of man's living condition. War was another factor in several ways since shortages were created and also machines for the needs of war. Industrial rivalry was another factor. These factors caused invention.

Horace Smith maintained that "a man was not free unless he was independent through knowing a trade or profession which earned him a living."

Daniel Wesson kept the lesson that "an apprentice who considers the master's business as his own and who is attentive and industrious will win the esteem of all acquaintances and will surely thrive in the world when he comes for himself."

Horace Smith and Daniel Wesson, though possessing a limited education, had revealed inventive genius and in the span of a lifetime through energy and skill had established a great internationally known industry as a veritable symbol of precision manufacture and workmanship.

XIII

1906-1953

1907—44 Hand Ejector—New Century
1911—22/32 Hand Ejector Target
1909—38 Double Action Perfected
1913—35 Automatic
1911—22/32 Heavy Duty Frame
1917—.455 Hand Ejector, Mark 2 English
1917—38 I Hand Ejector 3 (Regulation Police)
1917—45 Hand Ejector—U. S.
1925—22 Straight Line
1924—32 Automatic
1926—44 Target
1930—38/44 Outdoorsman
1931—K 22 Outdoorsman
1933—38/44 Military and Police
1937—22/32 Kit
1935—.357 Magnum
1936—38/22 Terrier
1938—.38 Military & Police
1940—Original K-22 Masterpiece
   New K-22 Masterpiece
1940—38 K 200
1942—38 Victory
1950—45 Army
1950—Model 45 Target
1950—38 Chiefs Special
1952—38 Centennial
1953—Airweight

1907—44 Model Hand Ejector

This model is also known as the New Century Model and the Triple Lock Model. It is similar in design to the Model .38 Hand Ejector or .38 Military and Police 1906 change. To withstand heavier breech pressure, a casing covered the forward cylinder yoke lock and extractor rod. It needed a wider frame and the cartridge recoil jammed the knuckle into the web between the thumb and the forefinger.

The barrel was made in 4, 5, 6½, and 7½ inches. It was a tapered round barrel, threaded, and held in position by cross dowel pin through frame. A lug raised from barrel body underside incasing the extractor rod and locking bolt.

The cylinder was 6 shot and grooved. The cylinder had inserted gas ring counter-bored to receive flanged head extractor with ratchet so designed as to allow hand to pass by tooth in carrying up.

The trigger was pivoted on stud screwed in frame. The locking bolt was in barrel lug that stuck out from face of barrel at thread shoulder and held in position by plunger and coil wire spring inserted in front end. The action of the closing yoke engaged the yoke cam latching in and locking yoke to barrel. It was released by the forward action of frame bolt engaging locking bolt through the medium of the center pin.

It was chambered for the .44 Russian cartridge and the cartridge was increased to 26 grains of black powder from 23 grains. The bullet was the same 246 grains and the case was lengthened.

In this first model 13,753 were manufactured for the 44; 1226 for 450 Eley; 21 for Colt 45, and during World War I, for the 450 Mark 2 British Service cartridge, 5000 were chambered.

In 1917 or with 15,525 serial number the casing covering of the front end of the extractor rod was removed because mud caked the casing and made it hard to work. This change was due to our World War I experience.

With 16,600 a heat-treated alloy steel cylinder was added and 727 were made for 45 Colt. The hammer shape at foot forward of rebound seat made to engage notch in trigger for increased double action throw. The knuckle above the grips at the back of the frame formed a sharp corner and in 1938 the grip was extended to the corner of the knuckle.

1908—22/32 Heavy Frame Target Revolver

This model may be called the "Bekeart Model" in honor of Phil Bekeart, a San Francisco firearms dealer who urged Smith & Wesson to produce a .22 that should be more than a miniature. He suggested something on the frame of the hand-ejector 32 of 1903. Smith & Wesson wanted some assurance that Bekeart's idea would be received by the public and Bekeart agreed to take one thousand. The order was filled and they were numbered in with the 32 Hand Ejector model. That is how this target revolver was born.

Chambered for 22 rim-fire, 22 short, 22 long and 22 long rifle, it was made in 6" barrel and was a six shooter. The hammer face was cut parallel to bolster to allow space to insert firing pin bushing to strike rim-fire cartridge on top.

The first thousand had a slide sight with round bead for flat face adjusting windage screw. Later it was changed to fl at slide dovetail shape lug for pointed adjusting windage screw.

1909—38 Double Action Perfected

This model was the latest change of the 38 Double Action 1 introduced in 1880. It had a heavier frame; the trigger guard was an integral part; the lock was changed to spiral springs; the hammer safeguarded, and a push button added.

This model was chambered for S & W center-fire cartridges. It was made in 3¼, 4, 5, 6-inch barrel lengths. This model was manufactured up to 1920.

1911—The 22/32 Target
22/32 Heavy Duty Frame—Hand Ejector

Serial numbers were run in with the .32 Hand Ejector Model of 1903 because it was modeled from that design. It is chambered for the .22 rim-fire, shot, long, long rifle and made as a Target revolver. It is 6 shot, only with 6-inch barrel, and length over all is 10½ inches. It weighs 23 ounces. The hammer face cut is parallel to bolster and allows space for firing pin bushing insert. This permits striking rim-fire cartridges on top. As Smith & Wesson explains this model, "Up to 1911, when we brought out this model, revolvers of 22 rim-fire caliber had all been too small and light to permit of really good shooting."

1913—Automatic .35
1924—Automatic .32

It was not until 1913 that Smith & Wesson entered the automatic pistol field. Colt had introduced their first model in 1900 based upon the John Browning patent of 1897 (U.S. 580,924). The automatic pistol was an immediate success. Between 1900 and 1912 one million automatic pistols of Browning design were made by Fabrique Nationale in Belgium. The Savage Arms Company, in addition to Colts, was an important manufacturer of the automatic pistol.

Smith & Wesson secured the United States rights to manufacture the automatic pistol developed by Charles P. Clement of Liege, Belgium. The barrel was 3½ inches. It was chambered for the 32 Automatic Colt Pistol Cartridge, a jacketed cartridge. Smith & Wesson produced the "half-mantle" cartridge. It had a cupro-nickle jacket that covered the tip and part of the surface. At the point of the bullet's largest diameter it was exposed lead.

The barrel of the automatic was made solid with the frame for accuracy. A bolt release catch to retract the breechblock and cock the pistol and not have to overcome the return spring tension was another feature. The barrel-frame assembly could be released and tilted for cleaning from the breech. A safety catch had to be pressed in order to fire. These were all good features but the automatic pistol did not prove successful and in January, 1920, its manufacture was discontinued.

In 1924 Smith & Wesson again tried to explore the capabilities of this field by the production of a 32 automatic pistol where the barrel was not hinged to the frame but was placed within a two-part slide. It therefore could be retracted separately from the forward part so that a cartridge could be loaded without compressing the recoil spring. This model was manufactured until 1937 and again Smith & Wesson abandoned the field.

1917—Model 45—Hand Ejector
Army Model 1917
1950 Army

The official U.S. Army hand gun in 1917 was Colt .45 automatic. At our entrance into World War I in 1917 the Army ordered Colt New Service and Smith & Wesson revolvers, but only if they could use the .45 government automatic ammunition. The .45 government automatic ammunition was rimless and Smith & Wesson extractors therefore could not grip them. This was solved with a semi-circular, 3-cartridge clip.

The 1917 model .45 was identical with the model .44, hand ejector, second model except for caliber and chambering and the barrel was 5½". The pitch of rifling was one twist to 14.569 inches and Colt was one turn to 16 inches. One hundred seventy-five thousand were manufactured.

After the war, a commercial model was developed in which the clips for cartridges were eliminated, a rim added to the cartridge, a lead jacket added and a bullet of 255 grains achieved.

In 1938 the Magna stock was added to lessen recoil.

The revolver is currently produced by Smith & Wesson and entitled, The 1950 Army. It has been brought up to date with a shortened action, lowered and broadened hammer spur, and in other small ways. It uses 45 A.C.P., 45 auto-rim, and 45 automatic wad cutter ammunition. Its barrel is standardized at 5½ inches; weighs 36¼ ounces; length over all 10 and ¾ inches; has a 1/10 inch service type front; square notch rear sights; and stocks are checked walnut with S. & W. monograms.

The Smith & Wesson catalog states that " it is a powerful and fast-handling gun for practical shooting within the range of 100 yards, with its most efficient area being considered from 25 to 50 yards. A rimmed cartridge known as the .45 Auto-Rim is available for this gun, but most users seem to prefer service ammunition and the faster loading which the clip makes possible. It should be here mentioned that while the semi-circular clips make for faster loading and extraction they are not necessary for the actual firing of the cartridge, since the cartridge is positioned by the forward end of the case which seats on a shoulder within the charge hole of the cylinder."

1917—Model 455 Mark 2, Hand Ejector, English Service

During World War I, Smith & Wesson manufactured 73,650 of this model for the British government. The first 5000 were made by reboring the New Century Model .44. The encasing frame was omitted. It has the barrel in 6½ inch only. Chambered for caliber 455 Mark 2, which is the Colt 455.

Fifty-nine thousand, one hundred fifty were manufactured for the English Service and 14,500 for the Canadian Service. The cylinder was rechambered, the hammer shape at the foot was changed to flat bearing engaging rebound slide, the knob on the extractor rod was made solid on rod with separate collar, the barrel lug that raised from barrel body that incased rod and yoke locking bolt removed, and the cut for the yoke cam on frame removed. The butt has swivel.

1917—38 Hand Ejector Regulation Police—1st Model

Designed as Model I Hand Ejector 32 Regulation Police with several variations. It is a five-chambered cylinder and available only in 4-inch barrel. Its length over all is 8½ inches and its weight 18 ounces. The model was designed principally for police work.

1923—22 Straight Line Single Shot Model

The object of this model was "to design an arm which would be easier for the average man to shoot well, and at the same time make several desirable improvements in the action."

It had a ten-inch barrel similar to the Olympic Model or Perfected Model Single-Shot or Third Model Single-Shot manufactured from 1909 but differed in many respects. It was heavier and opened by half-cocking and swinging the barrel to the right on a pivot just like some derringers.

The stock was akin to the S.&W. Clement automatic and Smith & Wesson's explanation for this is as follows: "We determined to design an arm which would be easier for the average man to shoot well, and at the same time make desirable improvements in the action. To this end we adapted the type of handle typical of the Auto Pistol, as when shot with the arm extended this shape makes the pistol point naturally without any bending of the wrist. Those who have learned to do deliberate shooting with a revolver or with our old model pistol will find this auto-shaped grip a very different proposition—but after a little practice, scores will begin to improve and shooting will be much easier, and require much less holding for each shot. For years expert pistol marksmen have criticized the swinging hammer and trigger on the ground that they tended to disturb the aim. To overcome this we have used a straight pull trigger and a hammer that moves in a straight line parallel to the bore."

This model was not well received.

1926—44 Target

This model is a heavy duty and large frame target revolver, 6 shot, 6½ inch barrel, 11¾ inches over-all and weighs 39½ ounces. It is chambered or 44 S.&W. Special and is heavy in order to counteract the heavy recoil of high-powered cartridges.

1930—38/44 Outdoorsman
(See 1933 38/44 Heavy Duty Revolver)

This is a target model with 6½ inch barrel chambered for the 38/44 Smith & Wesson Special, which is a high-speed cartridge. It is 6-shot, 11¾ inches length over-all and weighs 41¾ ounces. It is a heavy gun to counteract the heavy recoil of the cartridge.

1931—K-2 Outdoorsman

Cartridges had changed but not the revolvers to accommodate them. Remington-UMC designed a new 22 Long Rifle cartridge with a smokeless powder charge and a brass case. Potent with power the cartridge was fired but not enough of the revolver was left to fire another cartridge. Major Douglas Baird Wesson, grandson of Daniel, and Vice-President of Smith & Wesson, wanted to use this Hi-Speed cartridge and he packed the revolver cylinder with countersunk recesses to provide support for the cartridge rim. Lack of lateral support had caused the brass case to blow out around the rim where it joins the rest of the case.

Major Wesson then converted the cylinder and the barrel of the .38 Military & Police Model to use the .22 Hi-Speed by a countersunk or embedded chamber. The .22 Long Rifle is shorter than the .38, therefore it had to travel through the smooth-bore chamber before it contacted the barrel grooves. This decreased the maximum efficiency.

The model was produced until 1940 when an improved model known as the Original K-22 Masterpiece was introduced. This model was used until after World War II, at which time a new K-22 Masterpiece was manufactured.

1933—38/44 Heavy Duty

This is the service model of what originally was made as a target revolver—the 38/44 Outdoorsman.

Designed to combat criminal's bullet-proof vests and heavy metal automobile bodies, Major Wesson designed it for the .38 S.&W. Special cartridge. It is a high-speed cartridge and the forerunner of all modern .38 special high-speed cartridges. Designed by Elmer Keith and Major W. E. Witsil of Remington Arms Company and produced by Remington Arms Company in 1931.

A metal adapter was used to partly fill the space below the trigger guard to reduce the back drive. It is 6-shot, and has barrels in 4, 5 and 6½ inch lengths. With the 5-inch barrel it weighs 40 ounces.

Major Wesson realized that Smith & Wesson were lagging behind Colt. The .38 Superautomatic and the Colt automatic cartridges were being used by law enforcement officers and he realized the need for more potent ammunition and a powerful revolver. He designed this police arm by fitting a .44 frame with cylinder and barrel bored for .38.

Smith & Wesson describe this heavy duty weapon, "The action of this weapon has been shortened to achieve a very fast lock time, hammer spurs have been swaged to provide a firm grasp and make for fast handling. . . . On the underside of the barrel will be found a heavy lug which is an integral part of the barrel forging and which makes for a little more weight where it is needed to absorb recoil. This lug gives extra strength to the barrel and also acts as a housing to protect the extractor rod."

1935—The .22/32 Kit Gun

Chambered for a 22 Long Rifle, it is designed as round butt Regulation Police Model. It is a revolver, with a 4-inch barrel, 6-shot, and designed for the camper, the fisherman, the sportsman, or those who want a small, light and accurate gun. It weighs 21 ounces and has 1/10-inch Partridge or Pocket revolver front sight, square notch rear adjustment for windage and elevation.

1935—The .357 Magnum

The history of firearms is the story of ignition. Firearms began with the discovery of gunpowder and a gun shoots because the powder is set on fire.

Firearms did not come into general use in Europe until early in the fourteenth century, though primitive types of firearms were used by the Chinese as far back as 600 B.C.

Firearms were introduced in Europe as soon as Europe found out the formula for gunpowder. Roger Bacon is credited with its introduction into Europe in 1247. This knowledge of gunpowder made possible the firearm and revolutionized warfare. It helped the transition from the medieval age to the modern era. It hastened the end of the feudal system of the Middle Ages with its serfdom since the stone walls of the noblemens' castles were no longer impregnable.

Gunpowder was the propellant of the bullet. The black powder explosive was a combination of charcoal, sulphur and saltpeter that burned violently. Charcoal produced the gas volume, sulphur the temperature and saltpeter supplied the oxygen to cause fast burning. The frightening effect of the flame and the smoke of the first weapons were paramount to their lethal qualities, but gunpowder revolutionized warfare.

The five hundred years from 1350 to 1850 saw the evolution of systems of ignition. How to ignite the powder and how to produce the flash of ignition therefore was a vital development and a counterpart to the discovery of gunpowder. The earliest ignition of the gun was generally by means of a lighted wick applied to the touch-hole or vent in the breech. The fire flashed back to the main charge in the breech of the gun and discharged the bullet. We see the evolution of the systems of Serpentine ignition; Matchlock ignition; Wheel-lock ignition, Snaphaunce ignition, Flintlock ignition, Percussion ignition and Smith & Wesson's introduction of the modern Cartridge ignition.

In the nineteenth century we saw the greatest progress with the rapid technical improvements and increased efficiency of the revolver, the rifle, and the cartridge.

Then, having achieved the cartridge, began efforts to obtain greater velocities and flatter trajectories and greater accuracy. The balance between the weight of the gun, the length of the barrel, the twist of the rifling, and the cartridge were constantly being improved. The powder loads were increased and the cartridge cases lengthened until they were becoming unwieldy.

Prior to 1700 little attention was given to improvements in gunpowder but towards the end of the 1700's new substances were being experimented with by Lemery, Bayen, Fourcray, Vauquelin Lavoisier and Bethollet. The King of France in 1776 conducted a contest for the three best papers on the means of a quicker and cheaper manufacture of gunpowder. Berthollet in 1788 worked at the substitution of saltpeter by potassium chlorate. Later he tried to use argent fulminant but it detonated at the slightest shock, or at a sudden rise in temperature. Many experimented with various fulminates. In 1880 Howard, an Englishman, made a powder of saltpeter and mercurial fulminate. Lavoisier also suggested chlorate of potash be substituted for saltpeter, but with the fatal explosion of October 27, 1788, when LeTort, factory superintendent and Miss Chevraud, who was in Lavoisier's party of spectators, were killed, all further experimentation was dropped.

Black powder had incomplete combustion and this fouled the gun barrel. Black powder produced heavy smoke, which betrayed the position of the shooter. Black powder burned at a fast rate after ignition which meant that the projectile was ejected from the end of the barrel before it received the benefit of the full force of gas pressure and it also means that the breech had to be strong to absorb the impact. Since the projectile did not receive the full force of the gas pressure the result was a low rate of velocity and a high trajectory.

During the 1840's due to the combined discoveries of several, an improved propellant was developed—namely, smokeless powder. In 1845, Schoenbein, in Switzerland, produced guncotton by nitrating cellulose. In 1846, Sobero, in Italy, discovered nitroglycerin, formed by the action of nitric and sulphuric acids on glycerin. Here were two violent explosives. In 1855, Vieille, in France, discovered that if guncotton is dissolved in ether or alcohol and dried it is a propellant. Then came Alfred Noble and Sir Frederick Able to combine guncotton with nitroglycerin as a propellant and they could form gases without producing smoke.

The Vieille powder was known as single-base or bulk. The powder with nitroglycerin was known as double-base or dense. The single-base became available in the late 1880's and the double-base in the early 1890's.

Men who collect guns as a hobby are led by their interests into the subjects of metallurgy, ballistics, history and many other by-paths. The introduction of smokeless powder is one of the major innovations of the entire history of firearms. It presented a host of chemical, metallurgical and ballistical problems that have not yet been completely solved.

Smokeless powder burns cleaner; does not produce smoke; is by weight and volume more powerful than black powder; and the rate at which it is converted into gases can be controlled by regulating the different shapes and sizes of the grains and by coating the grains with graphite.

With smokeless powder it becomes possible to overcome the inertia of the projectile and push the projectile on its course by the reduction of the rate of combusion. Translated, it means a gradual pressure increase. A gradual pressure increase reduces the gas pressure on the breech, lessens the recoil, and increases the velocity, flattens the trajectory and improves the accuracy. This, in turn, means the greater powered smokeless powder needs smaller cartridges.

Looking backward at revolver ammunition we find the commonest types of the percussion period were the 31, 36 and 44, which became the 32, 38 and 44. In 1873 the 45 caliber center-fire cartridge was added. We must not forget the old reliable though a low-powered cartridge, the .22 rim-fire. Cartridge power varies according to size, bullet weight and powder capacity within these categories.

The development of ammunition led to the current smokeless powder and with non-corrosive primers. The non-corrosive primer leaves a residue upon discharge which is a preservative coating in the bore and the fired gun needs no cleaning.

From 1900 to 1928 smokeless powder cartridges were made with corrosive primers. Prior to 1900 we find only black powder so that we have witnessed rapid changes within a short span.

The most unusual of the 38 caliber cartridges developed is the Smith & Wesson "357 Magnum."

The highest-powered revolver ever manufactured was brought out in 1935. This handgun was born out of the new 357 Magnum cartridge which was produced through the combined efforts of Philip Sharpe and Merton Robinson of Winchester Repeating Arms Company.

The Magnum was specially developed and of materials required to handle the tremendous power. The barrel lengths varies from 3½ inches to 8¾ inches, since originally this gun was manufactured on custom orders only. Now barrel lengths are standardized at 3½, 5, 6, 6½ and 8⅜ inches. The actual bullet diameter is .38 S.&W. Special. It has six shots. The cylinder is heat-treated chrome-nickel steel. It has recessed head space and patented burnished chamber walls. The hammer is swaged spur deeply checked and notched for thumb tip clearance. Shortened hammer fall weighed and timed for uniform ignition. The trigger has S.&W. grooving, glass-hard point engaging hammer notch. The trigger pull in single action is three to four pounds, and 10 pounds in double-action. The frame has finely checked top strap matching barrel rib. Front and rear straps, S.&W. grooving. Sights: Choice of any S.&W. target sights, S.&W. Micrometer Click rear sights;

adjustable for windage and elevation. It is finished in either S.&W. bright blue or nickel. The ammunition is .357 S.&W. Magnum; .38 S.&W. Special Hi-Speed; .38 S.&W. Special; .38 S.&W. Special Mid-Range; and .38 Colt Special.

The 1952 catalog states: "Between 1400 and 1500 foot seconds muzzle velocity. Faster by far than the speeds heretofore obtained even by the small caliber light-weight bullets shot from foreign automatic pistols. Nearly 700 foot pounds muzzle energy. No other handgun and cartridge combination manufactured delivers anywhere near the speed and power or averages within hundreds of pounds of this terrific impact—and with this speed and power, accuracy. Machine rest tests have proven over and over again that the 6-shot groups fired at 25 yards can be covered with a dime, and machine rest groups comparable to those obtained with the finest grades of match target ammunition have been consistently produced. While its square-shouldered lead bullet will shoot through steel plates that are only dented by other cartridges that have heretofore been considered powerful, the Smith & Wesson .357 Magnum bullet will upset to approximately .50 caliber in 8 inches of soft paraffin where other bullets of lesser velocity pass through practically unchanged in form. In conducting penetration tests on wood boards it is always an astounding spectacle to witness the blasting and utter destruction of the wood fibers, and the condition of the recovered bullet tells of the power actually delivered.

"The Smith & Wesson 357 Magnum has far greater shock power than any 38, 44 or 45 we have ever tested, and with this power comes not only the precise grouping at close target ranges—at 100, 200, 500, 600, and even greater ranges the inherent power and accuracy continue to exist. At 550 yards it shows a penetration of 2 inches in live hickory, and the bullet is well upset.

"This cartridge, along with the Smith & Wesson 357 Magnum revolver in conjunction with which it was produced, has done much to hold organized crime under control. It would indeed be a mighty chore to supply body armor, bullet-proof glass, or otherwise equip the criminal and his car so as to be safe in the pursuance of his depredations. This combination of gun and cartridge continue to remain a valued asset to the highway patrolman, the special agent, and others who are responsible for the protection of the citizenry."

1936—38/32 2" Barrel Terrier

In conformance with a style trend toward snub-nosed revolvers, Smith & Wesson brought out the round butt Regulation Model with a two-inch barrel and chambered for .38 S.&W.; .38 Colt Police Positive; and S.&W. Super Police.

1938—38 Military & Police, Short Action, Round Butt

This model is a modification of the Military and Police square butt model made for law enforcement officers. It is 6-shot, chambered for .38 S.&W. Special, with 2, 4, 5, or 6-inch barrels and fixed sights.

In this model the design of the grip is stressed to provide a "saw handle" type of grip. Smith & Wesson makes three types—the Square, the Modified Square and the Round butt.

1940—Original K-22 Masterpiece

This model was a change of the K-22 first produced on January 3, 1931, in .22 rim-fire caliber and built on the frame of the .38 Military and Police. It had a somewhat shortened action, micrometer click sights, a built-in anti-backlash trigger and other improvements.

This model was discontinued after World War II in favor of new K-22 Masterpiece.

New K-22 Masterpiece

This model was introduced to replace its predecessor the ORIGINAL K-22 Masterpiece of 1940. It is a target gun and modeled on the frame of the .38 Military and Police revolver of which over three million have been sold.

Smith & Wesson's editorial blessing says:

"First, it has a ribbed barrel for more rigidity, and a little more weight out front. The front sight has a ramp base of which the front blade is an integral part, and the whole firmly pinned into the barrel rib. The bore is as usual precision reamed and rifled, after which it is lapped to a mirror-like finish, which is held to tenths of thousands of an inch in tolerance in size. The cylinder charge holes we continue to roll burnish for precision of diameter and ease of extraction. Hammers and triggers remain case-hardened for a lifetime of smooth and unchanging trigger pull. The rear sight is the same micrometer click sight which we first installed on our target models in 1940, and which has not seen the need of a single change in the last 12 years. The lockwork of this gun has been entirely redesigned so that this revolver is truly of the type known as a "short action" target weapon.

"The hammer fall has been reduced to the minimum requirements consistent with good ignition of the cartridge primer and maintenance of accuracy. The hammer spur on the standard K-22 has been swaged and formed so that it is easy to reach and the thumb will not slip off. The rearward action of the trigger is controlled by a built-in and fully adjustable trigger stop which prevents all rearward motion of the trigger after disengagement from the hammer notch. The standard stocks supplied with the K-22 are of walnut, Magna style, handsomely checkered to provide non-slipping qualities, and mounted with the Smith & Wesson monogram. The tangs, both front and rear, are grooved, as is the trigger, to prevent slipping in the handling of this weapon."

1940—K-200 (.38)

During World War II the British selected S.&W. Military and Police Model .38, which is a modification of the Military and Police Model of 1905 —4th change, and adapted it for the 38/200 Super Police. The .38 Super Police loads with a 200 grain bullet flattened at the point. The cylinder chambers were enlarged for the sharper taper. The choice was predicated on the theory that a heavy bullet moved by a light powder charge is more powerful than a medium weight bullet of the same diameter propelled by a heavier charge. This model has a 5-inch barrel.

1942—38 Special Victory Model

Smith & Wesson Victory model was the Military and Police model, 4-inch barrel, chambered for .38 Special and issued to the U.S. Naval Flying Forces, Coast Guard, and U.S. Maritime Commission.

Smith & Wesson also supplied the Federal Bureau of Investigation with a Military and Police 2-inch barrel revolver chambered for the .38 Special and known as the "Victory Model."

The 1950 Army

This model is 6-shot, 5½-inch barrel, length over-all is 10¾ inches and its weight is 36¼ ounces. The sights are 1/10 inch service type front; square notch rear.

The history of the weapon as related by Smith & Wesson is as follows:

"When America entered World War I, sidearms in great volume were needed and without delay. Remember that the service pistol shot the rimless A.C.P. cartridge, and it was out of the question to have more than one size of cartridge for each class of arms. It is a matter of history that Smith & Wesson solved the problem by designing and turning over to the government the semi-circular, 3-cartridge clip, which is used to this very day in revolvers chambered for the .45 A.C.P. cartridge.

"The old 1917 Army model was used by the American troops in World War I, and in the government test gave greater velocity, penetration, and accuracy than any other handgun tested. The 1950 Army model is the same weapon except that it has been brought up to date with a shortened action, lowered and broadened hammer spur, and in other small ways which make it a safer and more secure gun to handle. It is a powerful and fast-handling gun for practical shooting within the range of 100 yards with its most efficient area being considered from 25 to 50 yards. A rimmed cartridge known as the .45 Auto-Rim is available for this gun, but most users seem to prefer service ammunition and the faster loading which the clip makes possible.

"A precision rifled barrel, close chambering, perfect alignment, and hardened or heat-treated working parts, combine to resist the hammering of the heavy cartridge without shooting loose and without loss of accuracy."

1950 Model 45 Target

This is a target version of the military 1950 Army revolver. It is a six-shot, 6½-inch barrel, 11¾ inches over-all and weighs 39 ounces.

Smith & Wesson explains that "Target shooters have long been converting our military model for use in matches where .45 caliber was specified. They sought features not obtainable in the auto loaders, i.e., lighter and finer trigger pulls, positive function with various types of ammunition and others. To render a service to these sportsmen, we have developed a target version of the military model."

1950—38 Chiefs Special

This model derives its name from the fact that its first formal presentation was at the Conference of the International Association of Chiefs of Police, held at Colorado Springs in 1950.

The objective was to produce the smallest and lightest all-steel revolver chambered for the .38 S.&W. Special cartridge. It is 5-shot, barrel 2 inches, length over-all is 6½ inches; and weight is 19 ounces.

It was presented as a weapon for the law enforcement officer to either keep it exposed or concealed yet with adequate power. It has an external handle whereas the Centennial Model, though similar, is of hammerless design and for double action only.

1952—The Centennial

This gun "has been sired by two of the most successful and popular revolvers in the small gun field—the Safety Hammerless (or as it was catalogued, the "New Departure") model, which we produced from 1888 to World War II, and the Model J (or as it is herein catalogued the ".38 Chief's Special") which is currently receiving the most favorable notices in law enforcement circles."

It is a 5-shot, 2-inch barrel, 6½-inch length over-all and weighs 19 ounces. Its sights are fixed, 1/10 ramp type front; square notch rear. It is a hammerless gun designed for personal safety. It has the grip safety, or as it was called, a "squeeze safety" which consists of a movable bar running up and down the back of the grip which prevented the arm from being discharged unless the bar were held down or depressed into the butt of the gun during the act of firing.

Smith & Wesson recommends its use to the detective, or as an undercover weapon for the law enforcement officer. He will not need the use of the safety bar. They have therefore provided a special feature whereby the insertion of a simple pin in the mechanism blocks out the action of the safety bar. Another feature is an indicator provided on the top of the safety bar which remains a constant reminder as to whether the bar is operative or not. It has a white dot which indicates whether the weapon is safe or ready. If the white dot is in sight the safety bar is out, the mechanism of the gun is locked and the revolver is safe. With the white dot in sight, the trigger alone will not discharge the gun. The trigger will not function until the safety bar has been depressed. If no white dot can be seen, then the weapon is in ready condition and can be fired at will by merely pulling the trigger.

1953—Airweight Revolvers

Smith & Wesson is currently producing three models of special purpose revolvers in lightweight metal, known as the "Airweight" models. These guns are not intended to replace their regular line of all-steel revolvers, but "merely to supplement it."

These models are:
1. A special adaptation of the .38 Military and Police round butt model with 2" barrel, which will weigh 14⅜ ounces.
2. The Centennial model, which will weigh 11¼ ounces.
3. The .38 Chiefs Special, which will achieve a new low level of weight factor at 10¾ ounces.

At the same time Smith & Wesson looks to heavier large-frame revolvers and announces:

"Seemingly the trend today is towards a heavier handgun, the sheer weight contributing to steadier holding, a lessening of the recoil factor, and a faster recovery of the gun from the effects of recoil. Handgunners also are leaning towards revolvers which have a little extra weight in the forward part, feeling that the balance and hang of the gun is better.

"We have been through this same thing with our famous line of Masterpiece revolvers, and it has been our experience that these guns met with more popular acclaim with a few more ounces of weight added to the forward part of the weapon.

"By the same token, we have two of these large frame target revolvers which have been extremely popular guns in the past—the 38/44 Outdoorsman and the 1926 Model 44 Target. We are not particularly anxious to change these revolvers to any appreciable degree, if in so doing we will not add to them and make them even more desirable to those who seek out this type of gun. Accordingly, therefore, our progress in making our large-frame guns a little heavier is going to be very cautious, and we will start with the lightest one of the group first and see what the reaction of the shooters will be.

"As a feeler in this direction we have placed experimental models of the 1950 Model 45 in the hands of noted revolver men around the United States, for test and report. We are also preparing a production run of this model which is fitted with a heavier barrel and reaches the weight of approximately 43 ounces. If this gun meets with the enthusiasm which we expect to encounter, you may anticipate an expansion of this program to encompass our other large-frame target models."

This is indicative of the constant research and experiment and vigilance that Smith & Wesson is committed to in order to maintain a record that has earned them the reward of "Superior Revolvers Since 1852." President Carl Hellstrom says, "Ever mindful of our obligation to the public, we are constantly seeking to improve our products and facilities . . . we shall maintain ourselves as a dependable source of superior firearms for legitimate sports and protection of society."

In setting forth their policy of quality products of proven design, Smith & Wesson surveys their effort:

"All parts of any Smith & Wesson gun are interchangeable with corresponding parts of any Smith & Wesson gun of the same model. This is extremely important in Military and Police service under the strain of active duty.

"The major parts of our revolvers are drop forged. The working parts are tempered. . . The cylinders are machined from the finest of modern alloyed steels and heat treated to give a tensile strength of well over 100,000 pounds per square inch. Hammers and triggers are case-hardened to give notches, points and contact area a glass-hard surface over a shock resistant interior. . .

"On all military models a wide square-section front sight and flat-top square notch rear sight of corresponding size are supplied. This combination is a great improvement over the thin front and small U rear sight notch that was standard on all fixed sighted guns for many years, and it has been closely copied by other manufacturers. So much better shooting has been made possible by the square-section sights that, but for the matter of adjustment, they are equal to the Partridge sights regularly supplied on target models.

"Although in recent years the square butt has become the more popular, many still prefer the round butt because it is the less bulky and more easily carried. Consequently, many of our models, particularly of medium and small size, can be had with either square or round butt.

"In spite of the allure of plastics and other moulded materials now widely used in the making of gun stocks, we still believe that the most important requirement of a gun stock is to have a holding area which will not slip in the hands when moist from perspiration or tension. For that reason we will continue to furnish checkered stocks cut from selected grades of fine grained domestic walnut. While wooden stocks are far more costly to produce than moulded ones, the rich color and patina of Smith & Wesson stocks add a note of warmth and luxury that those made of artificial materials can never attain.

"The Magna stock... the particular virtue... is to distribute the recoil over the greatest possible area of the hand and to provide a comfortable and secure grasp of the arm. On target revolvers an oversized version of the Magna stock is provided which gives a somewhat fuller grip, and provides for a slightly lowered hold which seems preferred by the target shooters in the Timed and Rapid Fire sequences of their sport.

"Unless the trigger is pulled intentionally, it is impossible to discharge a modern Smith & Wesson revolver, as the hammer nose or firing pin is prevented from striking the cartridge primer by two independent and automatically operating safeties—the rebound slide upon which the lower end of the hammer rests, and the solid steel hammer block which lies between the hammer and the solid frame just below the hammer nose.

"Even if the arm should fall and strike on the hammer with sufficient force to mutilate or smash the hammer, these two safety devices make accidental discharge of the arm impossible. The firing pin can hit the cartridge primer only when the trigger is deliberately pulled back to its full-release position. No manipulation or slipping of the hammer during cocking can result in accidental firing at any time, since the falling path of the hammer will be mechanically blocked at all times except when the trigger is held in full release position.

"In the Smith & Wesson revolver the cylinder is rotated counter clockwise to permit the rotating mechanism and sideplate to be on the right side of the arm, away from the locking bolt and thumbpiece which must be placed on the left side and bolted through the frame.

"This arrangement allows larger parts and stronger action, and furthermore makes removal of the sideplate and cleaning and oiling a simple operation with no assembly of small parts necessary in order to restore the weapon to operative condition.

"In handguns, alignment of cylinder with barrel is all-important and it is absolutely essential that bore, charge hole, firing pin, center-pin, bolt, locking bolt, and yoke arm. all remain permanently in the same plane. Rigid inspection with range plugs which measure to 1/10,000 of an inch is the medium through which we keep barrels and cylinders in alignment. The cylinder rotates on a center-pin which enters the frame through a hardened bearing in the rear, and the forward end of the center rod is held in alignment by the locking bolt located in a lug on the under side of the barrel. Thus alignment of cylinder with frame and barrel is held between these two points of suspension with the yoke acting as a brace in the middle."

XIV

THE EPOCH

HORACE SMITH AND DANIEL BAIRD WESSON grew up in an age of profound revolutionary change. The period as the backdrop of their lives is worthy of note. Their lives and the period in which they live yield only little loose seeds; names, dates, facts and rumors. This narrative is composed only of these little things that have passed down the current of time. Yet they yield meaning and suggest far more than they articulate. They illumine little niches, but these little niche lights give substance to the entire wall.

The Industrial Revolution had come to America as early as the 1820's in the New England cotton industry. The jacquard loom had been introduced to this country in 1825, and in 1837 a power loom for making fancy cotton fabrics was invented by William Crompton, a process that was later to revolutionize the woolen industry. A series of new inventions had revolutionized industry.

Eli Whitney, who died in 1825, had already proclaimed that "a substitute for European skill must be sought in such an application of mechanism as to give all the regularity, accuracy and finish to the work which is there affected by a skill." He had already formulated the doctrine of precision concepts and techniques when he enunciated, "One of my primary objects is to form the tools so the tools themselves shall fashion the work and give to every part its just proportion—which when once accomplished, will give expedition, uniformity, and exactness to the whole. . . In short, the tools which I contemplate are similar to an engraving on copper plate from which may be taken a great number of impressions perceptibly alike."

It was the same Eli Whitney who invented the cotton gin that gave the institution of slavery a new lease in America, and started thereby a chain reaction of events that led to the War Between the States. The machine multiplies evil as well as good, depending on the use people make of it.

Whitney's principle of mass production succeeded because the time was ripe and because the government wanted quickly and cheaply a single item in great quantity. An invention must answer an immediate need.

America was defined by Carl Schurz as "a colony of mankind whose homeland is the world." It was these uprooted that were the American pioneers. They needed the firearm, the axe, and the plow. Joseph Jenks in 1646 was one of the earliest and one of the very few to be granted a patent in colonial Massachusetts. He gave the scythe its modern shape. Some of our founding fathers were inventors. Benjamin Franklin invented the stove, the bifocal and conducted electrical experiment. Jefferson invented a plow.

The conglomerate collection of European weapons brought by the immigrant became the Kentucky rifle of about 1725. It had a longer barrel and smaller bore. The longer barrel gave good holding weight, economized on powder, and the increased distance between the front and the rear sights increased the accuracy. The smaller bore gave higher velocity with less lead used. A lead ball was used to shoot with and it was placed in a patch of lubricated buckskin or cloth and it was rammed down the barrel. The patch helped clean the barrel of unburned black powder and lessened the loading time.

In 1832 about two-fifths of the New England factory workers were children. Compulsory education was being urged. But the social ills brought by the Industrial Revolution could be offset by going west—the frontier relieved economic pressure.

The Percussion Plains rifle evolved because the pioneer left the forested areas and moved into the plains and the mountains. Ignition had changed from flintlock to percussion. The need out on the plains was for a larger bore for the buffalo and a shorter barrel for ease of handling astride a horse.

Kentucky became a state in 1796, Ohio in 1803, Louisiana in 1812, Indiana, Mississippi, Illinois, Alabama, and Missouri between 1816 and 1821, Arkansas in 1836, and Michigan in 1837, Wisconsin and Iowa were added as territories in 1836. The nation was growing at a tremendous rate.

Transportation had new developments. Steamboats plied the Ohio and Mississippi since 1811 and on the Great Lakes since 1819. The Erie Canal opened in 1825. In 1817 the first transatlantic packets made regular monthly sailings. In 1827 the Baltimore and Ohio planned a line by means of horsepower and in 1829 the first steam locomotives were brought from England. Philadelphia by 1832 had built a canal and railroad to Pittsburgh. The National Road was finished as far as Columbus in 1833. Better transportation helped pour settlers westward.

The story of invention is part of the political and industrial revolution in America. Alexander Hamilton in his report on manufactures stated that "the defect of hands constitute the greatest obstacle to success" and to effect "an increase in hands, an accession of strength," he suggested water-powered machinery. Power that would drive machinery and machinery that would transform raw materials into finished products and finished products that would satisfy the insatiable needs and wants.

Urgent need answers with instant acceptance of an invention. The face of a continent was changing and the opening of the West was a causative agent for the metallic cartridge breech loader.

The first hot-air furnace was introduced in Worcester, Massachusetts, about 1835 when Daniel Wesson was ten.

Worcester, situated forty miles from Boston and sixty miles from Hartford, was a town of about thirty-six square miles, its main street, broad and planted with fine trees, extending about a mile from North to South. Its population in 1825, when Dan Wesson was born, was 3650. The Town Hall had just been completed. Stage coaches were almost on every road. Daily mail stage coaches went to Boston, Hartford and New York. The Erie Canal opened in 1825. Railroads were unknown.

The Latin grammar school on Thomas Street in Worcester was built in 1832 and Charles Thurber was elected headmaster. It was known as the Brick School House. It is possible that Dan Wesson attended this school afterwards.

As the far west frontier moved, California gold discovered, and with the introduction of percussion the demand became greater for a gun that could be loaded more rapidly and more easily, a practical repeating firearm that could be loaded at the breech. The breech-loader was a complicated and grotesque contraption with gas escape at the breech.

Samuel Colt had perfected a successful repeating revolver and he was becoming fabulously wealthy. But he had a basic patent and competition was closed. How could that be circumvented? The need was there. Premature inventions bring public apathy. The War Between the States accelerated the need.

The clouds of war were gathering at home and abroad with revolutions in Latin America. By the light of sperm oil, mechanics worked to develop a breech-loading repeating gun mass-produced.

The financial reward was substantial. Material progress has been furthered by competition for material reward and a Utopian non-competitive society would be dull and stagnant.

Samuel Colt had started manufacture of his firearms in Paterson, New Jersey, in 1836 but this firm failed. In 1847 Captain Walker of the Texas Rangers came to Samuel Colt with an order for the Colt revolver for use in the Mexican War. Samuel Colt made an arrangement with Eli Whitney, Jr., to produce the guns. The proceeds of this contract enabled Samuel Colt to start his own plant in Hartford.

Five years later Samuel Colt constructed his tremendous factory and was accumulating wealth at such a rapid pace that when he died in 1862 he left an estate of over five million dollars. The story of Colt's success touched the imagination of others—from poverty to riches, from obscurity to fame and so swiftly.

Colt was a disciple of Whitney's production methods and this system transformed production. It is the keystone of our technological progress. The gunsmith originally filled the individual order with individually fitted parts made by hand tools and none of the various parts interchangeable with other weapons. In 1792 the United States government authorized the creation of two national armories—Springfield and Harper's Ferry—and a few years thereafter began to subsidize private contractors. In 1798 Whitney received a government contract to make muskets. Eli Whitney invented, designed and built all of his machinery to manufacture interchangeable parts with machine tools. It was two years before he produced his first muskets copied from the French Charleville musket.

Our modern automobile industry received its precision techniques from Henry M. Leland, who, in 1906, took three Cadillacs to London and disassembled the cars, scrambled their parts and then reassembled three new automobiles. It was Whitney's principle of interchangeable parts carried to its heights by Samuel Colt. Henry M. Leland had worked at the Springfield Armory during the War Between the States and later worked at Colt's Armory. It is easy to trace that geneology. When Smith and Wesson started their plant in 1857 they utilized the advantages of interchangeable parts. Wesson was aware of its advantages from Robbins and Lawrence. They were among the few large sized firearms plants and a large plant was necessary to enjoy the benefits of interchangeable parts.

Henry Ford, who stands in the same line of descent that runs from American firearm manufacture defined it thusly:

"Mass production is not merely quantity production; for this may be had with none of the prerequisites of mass production, nor is it merely machine production, which may also exist without any resemblance to mass production. Mass production is the focusing upon a manufacturing project of the principles of power, accuracy, economy, system, continuity and speed. And the normal result is a productive organization that delivers in quantities a useful commodity of standard material, workmanship, and design at minimum cost."

The American standard of good living also flows from the same source—the standardization of machined parts. Smith and Wesson and their work may appear remote but in the sequel we find their contribution a significant one. It was the perseverance of Smith and Wesson added to the skill plus their inventive genius that makes their contribution a noteworthy one. They developed and achieved the successful culmination that revolutionized firearm ignition by metallic cartridge.

....

Smith and Wesson grew from a modest beginning into an international reputation as America grew up during that same century from 1850 to 1950.

Carl Reinhold Hellstrom, President of Smith & Wesson, is a native of Sweden, graduate of Tensta School at Uppsala, and received his degree of Civil Engineer at Christiana College in 1914. He came to the United States in 1916 as a member of the French Engineering Commission to the United States in 1916-17. Occupied the post of Assistant Superintendent of Allied Shell Production in the United States 1917-18; was discharged with honorary S.C.D. from U.S. Army in 1919 and was granted citizenship because of service in U.S. Army. He entered the field of industrial production in U.S. in 1919 and became identified with Smith & Wesson in 1931. He has been President of Smith & Wesson since 1946, when he was chosen by the Wesson family. He succeeded Harold Wesson, son of Frank and Sarah Lovell Wesson, who served as President from 1920 to 1946.

In 1950 Lafayette College conferred upon Hellstrom the honorary degree of Doctor of Science. Hellstrom is widely known for his broad technical experience as well as his administrative capacity. He has been and is an important factor in maintaining the record of Smith & Wesson's leadership. His father was a world-famous engineer who built the Trans-Siberian railroad.

"100 Years of Gunmaking" is the text of Carl Hellstrom's address delivered at the Smith & Wesson plant when Carl R. Hellstrom was guest of honor, October 21, 1952, of the Newcomen Society of England.

....

"S. & W." 100 Years of Gunmaking!
(1852-1952)
By CARL REINHOLD HELLSTROM
President of Smith & Wesson

While age in itself is a mark of distinction, the only real justification for recording time is purpose and accomplishment. This calls for a strict and accurate accounting of our space in time, and we hope that our report will justify your time and attention.

In reviewing the century of Smith and Wesson's existence, one becomes aware of the extraordinary changes which have taken place in the world and in our way of living. Much of this change was due, directly or indirectly, to the manufacturing techniques developed during the Nineteenth Century. Among the pioneers in gun development we find the names of Daniel Baird Wesson and Horace Smith. These men were impressed by the necessity for practical interchangeability of component parts in order to achieve volume production of quality merchandise. This now seems perfectly obvious, but we must not forget that in those days gunmaking was a highly specialized art, confined to small shops where the individual craftsman labored with the one object of completing each unit so that it would function properly.

The time was right to apply these new principles to the production of firearms. Our West still was frontier country where Indian wars continued relentlessly, and tension and unrest, lead ultimately to the Civil War, were already in evidence in 1852. Manufacture of firearms was becoming a thriving business, with scores of arms makers dotting Pennsylvania, New Jersey, New York, and the New England states. Though fully aware of this competition, Daniel Wesson and Horace Smith shared the conviction that a superior revolver, produced in accordance with their advanced ideas on manufacturing techniques, would outsell all competitive arms.

Both men came naturally by their mechanical skills, Daniel's father being a plow manufacturer in Worcester, Massachusetts, while Horace Smith's father, Silas, was a carpenter, who, in 1812, came to work at the Springfield Armory. Horace too worked in the Armory, spending eighteen years serving his apprenticeship and earning the designation of "master gun maker." Later he spent several months in the Whitney Armory as a toolmaker, before entering the employ of Oliver Allen of Norwich, Connecticut. It was in 1851, when he moved to Worcester to work for Allen and Luther, that he met Daniel Wesson.

Daniel at eighteen was working in his father's plow shop, while his brothers, Frank and Edwin, were engaged in making guns. Their father, Rufus Wesson, tried vigorously to discharge Daniel's interest in firearms, and to this end stipulated that he must either stay in the plow shop until he was twenty-one, or pay the sum of one hundred and fifty dollars for his release. Although this was a rather sizable sum to raise, Daniel managed it, and thereafter promptly joined his elder brother, Edwin, in Northboro, where he began his apprenticeship in the year 1843. When Edwin died, in 1850, young Daniel tried his wings in a partnership with a Thomas Warner, master Armorer, in Worchester. But this association didn't last. Young Daniel needed more experience. For the next two years he is hard to follow. First, he went to Grafton to help his brother, Frank, build a pistol; after which he spent some time in the shop of the Leonard Pistol Manfacturing Company in Charlestown, Massachusetts. In 1851, we find him at Allen and Luther's in Worcester. It was there that he and Horace Smith became friends.

Both of these men had by this time acquired considerable firsthand knowledge of contemporary arms makers, their products and processes. Moreover, due to his experience in the Springfield Armory, Horace Smith was familiar with all available types of explosives suitable for use in firearms.

Their first effort toward the development of a new type of handgun was a magazine type of arm, which, in its final form, became the volcanic pistol. In 1852, they formed the partnership of Smith and Wesson to manufacture this gun which they were convinced was an improvement over any existing model. They open a shop in Norwich, Connecticut, to produce their new model in calibers .30, .36 and .41. Their importance as a step in the development of modern arms is primarily due to the mechanical principles involved in the repeating mechanism, but the ammunition developed for the gun was also a departure from the prevailing muzzle-loading type.

The bullets used in the volcanic model were conical and of the hollow base type. The base cavity was so deep that the bullet might be described as a hollow bullet with an open base. This cavity was filled with a propellant powder which was held in place by a rubber seal and brass, or cork, closing disk. The anvil cup was of iron, and the edges notched to admit the flash of the primer to the powder charge.

The firing pin of the volcanic model was barbed, and in firing the gun the disk on the base of the bullet was forced onto the firing pin where it was held by the barb. When the finger lever retracted the firing pin, the base cap was stripped off and permitted to fall free of the gun.

The value of their development is attested by the fact that, in 1854, Smith and Wesson achieved their first gold medal, bestowed by the Maryland Institute of the Mechanical Arts for their patent repeating pistol exhibited at the Baltimore Exposition in that year.

The mechanical action of the volcanic pistol later was improved and applied to a repeating rifle developed by B. Tyler Henry of New Haven, who patented the famous "Henry Rifle." This gun was the forerunner of the now famous Winchester lever-action models. The patents covering this gun and bullet subsequently were sold to Oliver Winchester who organized the Volcanic Arms Company of New Haven, Connecticut, later to become the great Winchester Repeating Arms Company. Mr. Winchester prevailed upon Daniel Wesson to act as his plant superintendent until the new concern was well established.

In 1856, Daniel joined Horace Smith in Springfield where their new shop had been located on Market Street. Twenty-five men were employed, and production was commenced on a new type arm, even more revolutionary than the volcanic model. The first sales of this new gun were made in November 1857.

This gun is familiar to collectors as the Model Number 1, (First Model). It used the first breech loading rimfire cartridges of the present type, and was protected by patents issued to Rollin White, covering a cylinder with chambers bored clear through. All previous revolvers required the loading of chambers from the muzzle end; White's development permitted the manufacture of the first true breech loading hand arm. This revolutionary feature, together with the excellence of material and workmanship, brought immediate prosperity to the firm. Orders poured in faster than they could be filled, and the new shop was expanded and improved.

In 1860, a larger plant was built on Stockbridge Street, in Springfield, and during the War between the States six hundred men were employed. A .32 caliber rimfire revolver, of larger size and heavier weight, was added to the line in 1861; and while the Federal Government never procured the gun in volume as a military weapon, thousands were sold to soldiers who spent their own funds in order to be armed with the best available handgun.

The .32 caliber model was called the Model Number 2, and, while it was discontinued in 1865 when the Model 1½ was developed, it remained popular for years thereafter. Evidence of this is the fact that eleven years after the gun was discontinued, the famous "Wild Bill" Hickok was still carrying one when he was murdered in a Deadwood, South Dakota saloon.

It was inevitable that the new and popular Smith and Wesson revolvers would be imitated, and the young firm found early in the game that patents, like liberty, must be defended. Some of the arms built in violation of Smith and Wesson patents were turned over to the Company by the Courts.

By 1864, Smith and Wesson was convinced that a larger caliber was needed for the military, and proceeded to develop such an arm. This was the .44 American Model, first of the larger caliber metallic cartridge revolvers. It chambered a husky center-fire cartridge which performed very well, and added substantially to the prestige and prosperity of the Company. In a letter to Smith and Wesson from the famous Eli Whitney, dated August 24, 1864, Mr. Whitney remarked, "It seems a pity not to have your military arm bro't out during this war—it must be done very soon if at all." Apparently, the development was slow since the first production of the .44 American Model appeared in 1869. Undoubtedly, the usual wartime shortage of materials, labor, and machines, combined with the urgent demand for existent models, was responsible for the delay.

During the Summer of 1869, the Grand Duke Alexis of Russia visited the United States of America and was taken on a buffalo hunting trip by the famous "Buffalo Bill" Cody. Cody was armed with a Smith and Wesson .44 American Model revolver, and the Grand Duke returned to the East with a profound respect for Cody's skill and the performance of the heavy gun. In fact, he proceeded directly to Springfield to discuss the production of a slightly modified version for the Russian Army. An agreement was reached promptly (wonder of wonders!) and the .44 Smith and Wesson Russian model was born. During the next five years approximately a quarter million of these guns were made and shipped to the Russian Imperial Army.

Soon after production, a group of Russian "technicians" arrived in Springfield. Their number grew, and so did their curiosity concerning manufacturing techniques. The shop soon was swarming with overzealous Russians, and finally an exasperated Daniel Wesson called a halt and insisted that the "technicians" be returned to the land of the czars.

In 1874, Horace Smith sold his interest in the firm to D.B. Wesson, and retired. From that date the Wesson name became dominant. Daniel's sons, Walter H., Joseph H., and Frank L., were all taken into the firm. Walter Wesson was admitted as a partner in 1883, and his brother Joseph in 1887. The third brother, Frank L., was killed in a disastrous train wreck in 1887.

Production of the first Double Action revolver model had begun in February 1877, and while the original reception of this innovation was apathetic, its obvious advantages over the Single Action type were soon proven. Sales volume increased and approximately ninety-four thousand were made between 1880 and 1884, as compared to twenty-five thousand in the three previous years.

By 1911, over half a million .38 Double Action hinge frame revolvers had been built, but they were discontinued because of the increasing demand for the .38 Hand Ejector Military and Police Model, which had first been produced in 1899. The Military and Police, as the new gun was called, had the advantage of solid frame construction, which permitted use of the more powerful .38 Smith and Wesson Special cartridge.

The variety of calibers and models of revolvers increased tremendously during the closing years of the 19th Century and the early years of the Twentieth. Improvements in powders, priming, and cartridge metals made possible substantial strides in increased accuracy and power.

The quality of the product was most rigorously guarded by the Wessons, father and sons, and their pride in the excellence of their revolvers was shared by employees. There are still men in the shop who can remember when Daniel Wesson made frequent visits to the shipping room where he proceeded to open guns packed for shipment, and to subject them to a most critical inspection. Seldom did he find cause for complaint, but woe betide the inspector who permitted a defect—no matter how trivial—to escape him.

Aside from his duties in the manufacture of firearms, Mr. Wesson was active in many other fields. For a time he was President of the Cheney-Bigelow Wire Works; was one of the founders and a director of The First National Bank of Springfield. He built and endowed the Wesson Memorial and the Wesson Maternity Hospitals, and was a generous supporter of many charities. On August 4, 1906, Mr. Daniel Baird Wesson died, aged eighty-one. His passing marked the end of an era in the company's history.

Three years later, the firm was incorporated and Walter H. Wesson became President. Both he and his brother, Joseph, were particularly interested in machine tools, and were constantly acquiring new and improved equipment. Some of it was imported from Europe, and probably the first "backing-off" machine seen in this Country was purchased by them in Germany.

Improved equipment was reflected in better accuracy, and from 1900, the United States Olympic Pistol Team used the .22 caliber Smith and Wesson "Perfected" Single Shot pistol. The accuracy of this model is phenomenal and even current models cannot improve on targets obtained with it. Of interest to handgun shooters is the fact that, in 1910, Smith and Wesson produced a limited number of "Perfected" Single Shot pistols with barrels chambered short, so that the bullet had to be forced into the rifling in order to close the breech. This theoretically improved accuracy, but was somewhat risky as the cartridges sometimes were discharged while trying to force the gun shut. This gun never was produced as a regular commercial model, but is often referred to by collectors and shooters alike as the "Olympic" model.

Another famous Smith and Wesson, first produced in 1907, is the .44 Hand Ejector New Century model. This gun is more widely known as the "Triple Lock" because of a third device which secures the yoke by means of an extension of the locking bolt. This model was—and still is—very popular with shooters. It was discontinued in favor of a new military design early in the First World War because British and Canadian Ordnance men felt that the more complicated looking mechanism might be vulnerable to dirt or sand under rugged combat conditions.

This new military model was adapted for use by our own military forces during the First World War, in connection with the three-shot cartridge clip developed by Smith and Wesson to permit the use of rimless

.45 Automatic pistol cartridge in a revolver. The Federal Government requested permission from the company to have these clips produced by other concerns for the duration of the war, and millions were made prior to Armistice Day. Over three hundred thousand revolvers were built for use with this device during the war. This model is identified as the "Revolver, caliber .45, Model 1917." We recently have brought this gun up to date in design, and it is currently listed as the "1950 Model .45 Military."

Smith and Wesson's experiences during the war were hectic, and as a matter of fact, the factory was commandeered by the Federal Government on August 31, 1918. The plant was not restored to Smith and Wesson until January 31, 1919. The events leading up to the seizure of the factory by the Government had their origin in unrest created by professional agitators among the newer employees hired to meet wartime production goals. However, in spite of all the strife and turmoil of the times, Smith and Wesson succeeded in making a very genuine and much-needed contribution to America's armament. Twenty-four years later, thousands of these revolvers were still in service and highly prized by our fighting men in the Second World War.

During the First World War, Joseph Wesson was President, having succeeded his brother, Walter, in 1915. His nephews, Harold and Frank Wesson, were his junior officers. When he died, in the spring of 1920, the company's management fell to the third generation, with Harold Wesson as President, and Frank H. Wesson as Treasurer. Douglas and Victor, sons of Joseph H. Wesson, and a fifth grandson, Harcourt Wesson Bull, also were active in the business. Frank H. Wesson, now the only active survivor of the five grandsons, still serves as our Vice-President and Treasurer, having devoted over forty-seven years to the firm. Much of the sketch of his grandfather's earlier history has been made possible by his generous assistance in verifying, through family records, data not available in our archives. A great grandson of the founder, the son of Victor, Daniel Baird Wesson, is presently serving as an assistant superintendent in the plant.

For many years prior to the Second World War, restrictive legislation and the public's apparent apathy concerning the piecemeal encroachment on their Constitutional rights "to have and to bear arms" resulted in a shrinking market for guns. Under such circumstances, it is not surprising that the uncontrolled imports of foreign arms and the ruinous competition in the small arms industry, which began after the First World War, reduced several manufacturers to bankruptcy. It was, I think, due only to the loyalty and support of the Wesson family that this firm did not share the same fate. The rear guard action fought during the '20's did not admit maintenance of existing plant and equipment, much less replacement and modernization. The depression of the '30's further reduced the company's business to an economic liability, and ten years in the red did not recommend further extensions on the bond issue.

The beginning of the Second World War prompted a British order for automatic rifles. Anxious for some business, the company had agreed to redesign the rifle to accommodate a two grain step-up in the powder load of the ammunition, and secured a milion dollar advance on the order. The rifle failed to meet the rigid firing test with this high pressure, and, after rejection, the British demanded their money back under the original agreement.

Since the company already had expended $870,000 on the project, and consequently could not make the refund, a 10-day extension was granted for a new and final test. It was at the end of this test period that I was called into the plant as general superintendent; and it was at once clear to me that this job was the biggest challenge of my life. The only hopeful sign in the situation was the apparent frankness and determination of these descendants of a pioneer to do or die. They all freely admitted their mistake in not confining themselves to their specialty, which was side arms; and, upon the final rejection of the rifle, it was decided to concentrate on building up the company's business in this field.

As a result of the Dunkirk evacuation, the company had received substantial orders for revolvers from the various British Dominions, and England herself was sorely in need of a large quantity of this type of arms. Arrangements were made to pay back the British advance by supplying revolvers at a fixed rate, and ways had to be found to produce guns in quantities heretofore undreamed of. At the same time, production economies had to be effected under the fixed price contract in order to meet the payments.

Our success in pulling out of this desperate period is a monument to cooperative efforts among men. Each man in the plant seemed to accept the situation as a personal challenge to his ability and ingenuity. Every possible piece of equipment was patched and repaired, to avoid the expense of machine tool replacements. Nearly two thousand of these machine tools were over fifty years old, and a good percentage had been the backbone of Smith and Wesson's Civil War production.

The obstacles to normal production seemed insurmountable at first. We had no drawings or tool catalogs, all work being done to templates or models. No fixed tolerances could be established, and all threads and tapers were non-standard, having been established long before the day of National Standards governing such details. As a result, we were forced to make our own tools, and we were, I believe, at that time the most self-sufficient plant in the country.

Some time after the United States entered the war, an accident, resulting in the death of a sailor, brought to focus the danger of issuing side arms to raw recruits. A loaded revolver had been dropped on the deck of a battleship, discharging a cartridge. At an ordnance meeting the following day, in reviewing the long history of accidents, an urgent demand was made upon us to design, within three days, some positive safety device superior to those in use up to that time. It was during the three hectic days and nights which followed that our revolvers became the safest side arms in the world. How many lives have actually been saved by our new hammer block, we don't know, but, since that time, we have not heard of any such accident from our military establishments, or any other source.

The small arms business during the first part of the Second World War was, of course, highly competitive, and to meet this competition we had to redesign the toolings of our .38 Military guns for more efficient and economical quantity production. These engineering changes proved very successful. Our total contribution to the war effort reached the incredible total of one million, three hundred eleven thousand revolvers, representing over 80 percent of the total Allied requirements.

While our gross profits, as a result of these forced war operations, were running into millions, the company could retain very little of these profits for itself. Government renegotiations took much of the fruit of our technological developments, and the 80 percent wartime excess profits tax was, in our case, almost confiscatory. It should be remembered that wartime earnings were restricted in relation to invested capital, and practically all of the Smith and Wesson facilities had been written off under standard amortization years before.

It soon became evident that our nearly 100-year old facilities on Stockbridge Street in Springfield could not stand a production load of twelve hundred guns per day. The old plant consisted of twenty-one mill type multistoried buildings, located in the heart of Springfield. Our drop forge hammers running twenty-four hours a day set up a vibration felt for blocks around, and, for the safety of the plant, it was imperative to find another location for the forge shop.

A complete survey of all available industrial areas inside the perimeter of our labor pool produced nothing within our economic reach. We did discover a 126-acre swamp West of Roosevelt Avenue, in Spring, consisting of several parcels, most of which had been abandoned for taxes. Adjacent to this swamp were two high hills of suitable fill, and since the area bordered on two railroads, a park, and a golf course, and was only four miles from our plant, we felt it was worth looking into.

Investigation disclosed that nine feet below the swamp surface there was (with the exception of some sink holes and quicksand areas) a solid sand-clay stratum. A rough calculation of the hills revealed approximately one million, one hundred thousand yards of available fill which would be sufficient to bring the area up to acceptable grade, providing we could lower the water table three feet. According to existing contour maps the topography showed this to be possible, and a check run to existing bench marks confirmed a satisfactory sewer drop from the area. The watersheds to the swamp, of course, presented a problem as to surface drainage, but the five mile even grade to the Connecticut River seemed to assure a satisfactory solution.

The site eventually was bought in 1942 at a total cost of $29,500, and a bulldozer and other earth-moving equipment were acquired and put to work. Through certificates of necessity we were able to erect the drop forge, warehouse, and boiler room buildings; and, under a $50,000 land improvement certificate, some drainage, a railroad siding, and temporary roads were added. That was all that our so-called war profit would allow, and aside from surveys and layouts (beyond War Production Board jurisdiction, of course!) the project was dormant until October, 1946.

These additional facilities did, however, enable us to carry out all of our contract commitments with the Armed Services. We voluntarily had suspended all civilian production after Dunkirk, and we take pride in the fact that every single revolver delivery schedule set up for us by the United States or Allied governments was completed in full, and on time.

The war took much of our strength. When it was over, I believe we were all exhausted from our efforts. Harold Wesson never fully recovered, and, in August, 1946, this man, who had guided the firm for twenty-five eventful years, died. Nothing, save my promise to this harassed and worried friend, would have induced me to accept the burdens of his office.

V-J Day found the company straddled between its old factory building on Stockbridge Street and the new drop forge and warehouse facilities on

Roosevelt Avenue. We were faced with a conversion to peacetime operation with wornout machinery and dissipated toolings. Contract terminations did leave us an unbalanced process inventory for military model guns, but nothing to start on for the civilian market. Our balance sheet at the end of the fiscal year, July 31, 1946, showed cash of $889,000 and receivables and claims for current tax adjustments of $246,000. Our liabilities were listed at $165,000. The receivables, however, had been frozen by a revenue agent's disallowance of claims made by the company in prior years. The report of examination, to August, 1944, claimed tax deficiencies of nearly two million dollars, due principally to commissions and refunds paid under contracts on foreign business, and a relief claim under Code Section 721, which is still pending before the tax court.

We had readjusted our working force to some twelve hundred employees after the war, but to operate under the uncertainty of civilian receivables with a working capital of $700. per employee was altogether different from the assured wartime payments of the Government. There could be no definite appropriations for anything. The best we could get from the stockholders' and directors' meetings was the authority to "go ahead and do the best you can." How we did it, I really don't know!

It seemed that our struggle had just begun. In order to recapture and maintain ourselves in the civilian markets, our entire line had to be redesigned and tooled to modern standards. New models had to be developed and made ready to meet competition. Foreign outlets and representations, dormant for years, had to be re-established all over the world. A swamp had to be drained and filled, three hundred and fifty thousand feet of factory built, and over two thousand belt-driven tools had to be rebuilt and converted to individual motor drives. And last, but not least, we had to maintain continuity of operations while moving the entire plant to the new location.

We, of course, could not afford to hire engineers and architects to plan our factory. This we had to do ourselves in our spare time. Except for the drop forge building, there never was money enough in sight to let a contract for any one unit. The plant, as you see it, was constructed by daywork labor supplied by the contracting firm of Ernest F. Carlson Company as, and when, we made enough money to pay them. It was our policy not to borrow any money (as if we could!) for the company rehabilitation. Had it not been for the facilities, patience, and able assistance of this firm, we could not have held the over-all cost within our optimistic estimate of six dollars per foot. The plant was completed in August, 1949.

It should be remembered that right here we are sitting on top of fifteen feet of fill, and yet, as you see, we are still fifteen feet below grade, which at the upper yard level increases to twenty feet. It is obvious then, that the described, otherwise costly arrangements, were ours for little more than planning costs before filling operations began. The plant is floating on a clay stratum, supported by skin friction piling, and walls below grade are of counterfort construction, heavily reinforced; the forts carry a twenty-foot sand cushion, and all areas within one hundred feet of the plant have sand fill from footing levels.

As an arms manufacturer, we can expect to be a prime target in case of war. The size of the plant bears no relation to its importance. If you consider that our heaviest complete unit weighs forty-seven ounces and each such unit has an average of seventy-two precision parts, some with tolerances less than one thousandth of an inch, you can form some idea of our relative size.

Each complete gun requires over two thousand separate operations, and there are over five hundred inspections of parts and components before a gun goes to test firing. Hundreds of specially built machine tools, dies, gauges, and other irreplaceable equipment are involved. It was, therefore, our patriotic duty to plan for the safety of our employees and equipment in order to insure a dependable supply of weapons for our Armed Forces in case of trouble.

While we cannot, for security reasons, discuss some of the plant's features, it is obvious that the five hundred foot setback of the plant from fencelines, armed guards, electrically controlled gates, and an adverse focus floodlight system, leave little opportunity for unauthorized approach. Transformers, dust collectors, elevator control panels, and all service lines are below grade and immune to strafing or incendiary missiles. The open end courts will, of course, minimize cushion damage in case of aerial bombing.

Although we currently are using public utilities, we have our own standby facilities, such as a well distributed septic tank system with leach beds of sufficient size to operate indefinitely. We have our own water reservoir of ten million gallons, with driven wells for pure water. We have a bank of high pressure boilers of sufficient capacity to produce one thousand K.V.A.'s through a standby turbine-generator. Oil storage is distributed in underground tanks, none of which is over one hundred and twenty-five thousand gallons capacity, and there are dual ventilating and telephone systems.

High tension power distribution to the various underground transformer vaults are all below a twenty-foot sand cushion. All transformers are connected in series, and of sufficient size to carry the load in case of failure of any individual unit. Low tension distribution is through a plug duct system to facilitate rapid moving of machinery to underground areas in case of air raids.

Sufficient underground food storage, hospital, rest rooms, and toilet facilities are provided for our workers, and we could, I believe, maintain ourselves in comparative safety and carry on limited production for a considerable time. A direct hit by a heavy bomb at the right place would undoubtedly penetrate a compartment, but could do little damage to the rest of the area. As to the atomic or hydrogen bomb,—well, we are not responsible for the extravagant claims made by feature writers, and under a direct hit—who cares!

As to cost and maintenance, we are far ahead of the game. Saving in copper alone paid for the extra transformers, and constant jacket temperatures are worth the cost of the vaults. Our boilers in peacetime supply our heating, and process steam at low pressure, and the B.T.U. efficiency is saving their cost in fuel. The tunnel system relieves interdepartmental traffic, and makes possible the establishment of special security areas in wartime. It facilitates flexible line production of several models at the same time during normal operations. The underground areas are convenient for storage and for all sorts of extra activities. They made feasible exposed service line installations throughout the plant, as well as a standby gravity condensate system in case of pump or power failures.

The heating and ventilating costs for the below grade areas are practically nil. Our compressor bank takes nearly fifty thousand cubic feet of air per minute and is connected to an intake mixing chamber with shaft extending above the roof. This is equipped with a hydrostatically controlled free air louver and connected by ducts to all underground areas. This gives us a controlled draft-free suction arrangement effecting continual changes of heated air throughout both levels of the plant. Special shafts in fire walls, carried above roof level, admit pure air as needed. These shaft outlets have filter frames for use in case of gas attacks. The hydrostat would then insure an even oxygen supply to the monoxide-laden breathing air, being removed from this underground area through the mixing chamber. The air would otherwise affect the combustion efficiency of our furnaces and upset the pyrometric controls. During shut-downs or power failures, natural draft currents operate the system in reverse.

The gains from tooling and engineering changes enabled us to supply over a million guns for the war effort at a cost of 20 percent below that of our competitors. The economies and capacities of this new plant have made possible our maintaining a stable price for our product. In face of constantly increasing labor and material costs, we have been able to absorb all but a 10 percent increase in the cost of guns since 1946, yet our earnings during those years have been the largest in our history. In the last decade, we have made more arms than in the entire previous existence of the company, and our capacity right now is larger than that of the combined revolver manufacturers of the world. Were you to travel around the world, I don't believe you would find a country where you could not be under the protection of a Smith and Wesson gun. A policeman in New York, San Francisco, Bangkok, Ankara, Capetown, or Rio de Janeiro, will usually have one of our revolvers in his holster. A Canadian Mountie, a Border Highway Patrol Officer, a Texas Ranger, or a Cabinero will give you dependable protection with a Smith and Wesson revolver. Or should you be a target shooter, you will find our target models on the range wherever championship matches are held. If you are a sportsman or hunter, you will find a suitable Smith and Wesson whether you hunt moose in Alaska or go on an African safari. Our guns were the trusted friends of soldiers on every battlefront during the last war, and they are found today in every military establishment of the Allied world. Thousands of our Combat Masterpiece and Magnum revolvers, while not being regulation issue, have been privately purchased by our soldiers everywhere.

Ours is a world-wide business, but so are our problems. Economic changes, unstable currencies, import and export restrictions, punitive regulations, and changing requirements do not permit complacency. Lighter weapons, new explosives, and higher pressures, with their attendant problems of interior ballistics, are compelling reasons for strict metallurgical and process controls and continual mechanical research. Gone are the days of heavy guns and slow burning powders. Today, we place in the hands of the shooter a few ounces of metal, holding a cartridge developing pressures up to forty thousand pounds—ten times more powerful than the revolutionary volcanic pistol of one hundred years ago. Gone, too, are the free and simple methods of doing business, and the possibility of sufficient cushions against a depression cycle.

But we have greatly expended our line, we have a wider earning base, and are carrying forward an extensive research and development program for diversification of our business. We have a new plant with ample space for expansion, and an organization ready to be of service to our country, come peace or war.

These are our accomplishments by which we justify our age. Our employees, our stockholders, and the management are united in a common purpose to continue in our tradition of craftsmanship and integrity.

Cal.	Year	Model	Cyl.	Barrel	Serial Nos.	Patent Dates	Misc.	Price
22 r.f.	1857-60	1-1	7 plain	3³⁄₁₆ oct.	1-58,000	Apr. 3, '55; July 5, '59; Dec. 18, '60	Jointed thumb piece hammer; brass nickel plate frame; rd. side plate	45.00
	1860-67	1-2	7 plain	3³⁄₁₆" oct.	58,001-126,430		Solid thumb piece, hammer with lug; steel frame; irregular side plate	35.00
	1868-79	1-3	7 fluted	3 rd.	1-128,538		Fluted cyl; rd. bbl.	25.00
22 r.f.	1893-1905	S.S.1	S.S.	6, 8, 10	Numbered with S.A. 38		Flanged thumbpiece; 3,198 mfgd. Also available in 32 & 38.	35.00
	1905-1909	S.S.2	S.S.	6, 8, 10	1-4617		Only target gun. No holster flanges; no hand and stop slots.	35.00
	1909-1923	S.S.3	S.S.	6, 8, 10	4618-7305		"Perfected Target Pistol." Uses 32 H.E. 1903 parts.	35.00
22 r.f.; s.,l., or l.r.	1902-1906	M.H.E.-1	7	2¼, 3, 3½	1-4575		Hand Ejector; Bbl. screws to frame.	50.00
	1906	M.H.E.-2	7	3, 3½	4576-13,950		No rear end cyl. bolt	50.00
	1911	M.H.E.-3	7	3, 3½	13,951-26,154		32 Perfected Hand Ejector; Notched trigger full throw d.a.	50.00
22	1923-1930	S.S. straight line	S.S.					35.00
32 r.f.	1861-1874	Old No. 2	6	5, 6" oct.	76,502 mfgd.	Apr. 3, 1855, July 5, 1859, Dec. 18, 1860.	Though not officially adopted but used as Army Revolver used in War between States.	30.00

Cal.	Year	Model	Cyl.	Barrel	Serial Nos.	Patent Dates	Misc.	Price
32 r.f.	1863	Patent infringement action in Mass. Federal Court and the following mfgrs. They were die-stamped, "Manufactured for Smith and Wesson." Warner delivered 1513 (1863) Pond " 4486 (1863-4) Moore " 3376 (1863) Lowell " 8682 (1865-72)						
32 r.f.	1865–1869	Old 1½	5 plain	3½" oct.	25,224 mfgd. and nos. continue from Old No. 2			20.00
	1869–1875	New 1¼	5 grooves	3½" rd.	93,980 mfgd.	Apr. 3, 1855 July 5, 1859 Nov. 21, 1866		18.00
32 S&W c.f.	1878–1892	S.A.-1½	5	3, 3½, 6	1–97,900	Jan. 17, 24, '65; July 11, '65; Aug. 24, '69; Feb. 20, '77; Reissue July 25, '71; Apr. 20, '75; Dec. 18, '77	Stock first used was plain walnut later changed to checked rubber. Jointed catch on first extractor lifter; later sliding catch.	18.00
32 S&W c.f.	1880	32 D.A. -1 -2	5 5	3" 3, 3½	1–22,172 22,173–43,405	Reissue July 25, '71; May 11, '80; Jan. 3, '82	Cyl. grooves lengthened; rocker stop, freeing cuts and notches removed	20.00 17.00
	1919	-3	5	3, 3½, 6	43,406–327,641		Stop and rear sear changed at point of engagement; fully encased in trigger slot.	17.00
32 S&W c.f.	1888–1900	32 Safety -1	5	3, 3½	1–91,417	Sept. 11, '83; Oct. 2, '83; Aug. 4, '85	Catalogued as "Safety Hammerless," "New Departure." Flat leaf latch spring changed to coil wire spring with 53,400.	18.00

Cal.	Year	Model	Cyl.	Barrel	Serial Nos.	Patent Dates	Misc.	Price
32 S&W c.r. S&W Short	1900-1922	-2	5	2, 3"	91-418 226,880		Latch spring reverted back to flat leaf	20.00
32 c.f. S&W S&W long	1896-	Model 1 Hand Ejector 1	6	3¼, 4¼, 6	1-19,712	July 1, '84; Apr. 9, '89; Mar. 27, '94; May 29, '94; May 21, '95; July 16, '95	After 1st 1,000 a large head screw added to hold plate at hammer cut.	25.00
32 c.f. S&W. S&W.l.	1903	1903 Hand Ejector 1903 H.E.-1 1903 H.E.-2 1903 H.E.-3 1903 H.E.-4 1903 H.E.-5	6	3½, 4¼, 6 3¼, 4¼, 6	1-19,425 19,426-51 51-127- 95,500 95,501- 96,125 96,126- 102,500 102,501- 264,856	Apr. 9, '69; Mar. 27, '94 May 21, '95; Aug. 4, '96; Dec. 22, '96; Oct. 4, '98; Oct. 8, '01; Sept. 2, '02	Chafing pin added to trigger; cyl. stop notches elongated. Square wire coil trigger spring in rebound slide Patent date Feb. 6, '06 added. Minor changes Before hammer release for fall, hand slot allows stop to latch Chafing bushings removed from hammer; sear lengthened. Trigger omits hand pin slot, hand lever pin hole, hand lever spring hole.	35.00 35.00 35.00 35.00 35.00 35.00 35.00
		Regulation Police			331,320-		"Hand Ejector-3 issue." Same as 1903 H.E. 6 except frame tang has shoulder at back strap.	35.00

Cal.	Year	Model	Cyl.	Barrel	Serial Nos.	Patent Dates	Misc.	Price
32/20 W	1899	32/20 H.E.-1	6	4, 5, 6, 6½	1-5, 311	July 1, '84; Apr. 9, '89	Winchester produced 32 cal. 20 gr. powder cartridge & S.W. made revolver for same.	35.00
		32/20 H.E.1902 1902-1			5,312- 9,811 9,812- 18,125	May 21, '95; July 16, '95 Aug. 4, '96; Dec. 22, '96		30.00
		1905			18,126- 22,426	Oct. 4, '98	Oct. 8, '01; Dec. 17, '01 added patent dates	30.00
		1905-1			22,427- 45,200			30.00
		1905-2&3			45,201- 65,700	Dec. 17, '01	Added pat. date Feb. 6, '06 Sept. 14, '09; patent date added.	30.00
	1924- 1942	1905-4			65,701- 127,160-144,684		Dec. 29, '14 patent date added.	30.00
38 S&W c.f.	1876	S.A.2-1	5	3¼, 4, 5	1-24,633	Jan. 17, 24, '65; July 11, '65	Extractor gear catch cut is in bottom strat rear of barrel joint.	25.00
		S.A.2-2	5		24,634- 108,255	Aug. 24, '69; Jan. 19, '75 Reissue	Extractor gear catch cut eliminated. Sliding cam replaces ext. gear.	25.00
	1891 1891- 1911	S.A.1891 -3	5	3¼, 4, 5, 6	1-28,107	July 25, '71 Apr. 20, '75; Feb. 18-20, '77; May 11, '80	Model 1891 stamped on top strap. About 7,000 made for Russian trade with straight instead of bow shape guard & half cock notch of hammer omitted.	25.00

Cal.	Year	Model	Cyl.	Barrel	Serial Nos.	Patent Dates	Misc.	Price
38 S&W c.f.	1880	D.A.-1	5	3¼	1-25,000	Jan. 17-24, '65; July 11, '65	Side plate cuts run across side of frame.	20.00
		D.A.-2	5	3¼	25,000-119,000	Aug. 24, '69; Jan. 19, '75 Reissue	Side plate cut made with irregular edges to strengthen side walls.	20.00
		D.A.-3	5	3¼	119,000-332,700	July 25, '71	Rear sear latch notch moved to increase leverage and reduce pull.	20.00
		D.A.-4	5	3¼, 4, 5, 6	332,701-539,300	May 11, '80; Jan. 3, '82; Apr. 9, '89	Side walls extended to cover exposed action of rear sear	25.00
	1911	D.A.-5	5	3¼, 4, 5, 6	539,301-554,077			25.00
	1902-1920	D.A. Perfected	5	3¼, 4, 5, 6	1-58,398	Aug. 4, '96; Dec. 22, '96; Oct. 8, '01; Feb. 6, '06; Sept. 14, '09		40.00
38 S&W c.f.	1887	Safety -1	5	3¼, 4, 5	1-5,000	Feb. 2, '77; Dec. 18, '77;	"Safety Hammerless" "New Departure"	30.00
		-2	5	3¼, 4, 5	5,000-42,483	May 11, '80; Sept. 11, '83; Oct. 2, '83 Aug. 4, '85	Also pat. dates Feb. 14, '86; Apr. 9, '89; June 3, '90	30.00
		-3	5	3¼, 4, 5, 6	42,484-116,002		Hammer stop changed as precaution against accidental discharge.	30.00
		-4	5	3¼, 4, 5, 6	116,003-190,064			30.00
		-5	5	3¼, 4, 5, 6	190,065-			30.00
38 long Colt	1899-	Hand Ejector-1	6	4, 5, 6, 6½	1-20,975	July 1, '84; Apr. 9, '69; May 21, '95; July 16, '95; Aug. 4, '96; Dec. 22, '96; Oct. 4, '98	2,000 mfgd. for U. S. Navy; 1,000 for U. S. Army, 6½" bbl. "Military and Police Model." 5" bbl. is Police Model. (Military) Navy marked "S&W Navy 55.00 1899 U.S.N.";–"U.S. (All Others) Army Model 1899" 40.00	

The Story of the Revolver

Cal.	Year	Model	Cyl.	Barrel	Serial Nos.	Patent Dates	Misc.	Price
38 S&W Spec. U.S. Service		Hand Ejector-1902	6	4, 5, 6, 6½	20,976-33,803	Oct. 8, '01; Dec. 17, '01	Cyl. stop notched lined with shims, Extractor rod dia. increased;	35.00
		Hand Ejector-1902-03	6	4, 5, 6, 6½	33,804-62,449		Tang changed for square butt stock, 1904, about 58,000	30.00
		Hand Ejector-1905	6	4, 5, 6, 6½	62,450-73,250		Cyl. stop notches elongated plus minor changes	30.00
		Hand Ejector 1.1905-06	6	4, 5, 6, 6½	73,251-146,899	Feb. 6, '06	Rib added at hammer seat; extractor doweled.	30.00
		Hand Ejector 2.1905-06	6	4, 5, 6, 6½				30.00
		Hand Ejector 3.1905-09	6	4, 5, 6, 6½	146,900-241,703	Sept. 14, '09	Sear lengthened to engage trigger	30.00
	1924	Hand Ejector 4.1905	6	4, 5, 6, 6½	241,704-497,509 "C" Series, Third Million	Dec. 29, '14	Removed chafing pin and omitted hand pin slot of trigger.	30.00
38 S&W	1917 / 1924-1953	Hand Ejector-1 Regulation Police Still in Production	5	4"	1-28,465	Feb. 6, '08; Sept. 14, '09; Dec. 29, '14	Designed as Model I hand ejector, 32 regulation police with minor changes.	40.00
44S.A. American 44S&W American c.f.	1871	3-1 Army Revolver	6	8"		July 10, 1860; Jan. 17, 1865; Feb. 17, 1865; July 11, 1865; Aug. 24, 1869	Finished in blue or nickel. Square butt. 1,000 mfgd. for U.S. Army. Awarded gold medal 1873 International Exhibition.	45.00

Cal.	Year	Model	Cyl.	Barrel	Serial Nos.	Patent Dates	Misc.	Price
44 S.A. American		3-2	6	8"		Same as 3-1	Designed as 3-1 except that the cut across the face above the hammer nose for the barrel-catch lock was eliminated; and the locking lug extending from the body-engaging hammer on the barrel catch was removed.	38.00
44 Turkish r.f.	1879	3. Turkish	6	6½"		Same as 3-1	About 5,000 mfgd. Designed as 3-1 except bbl. length; Hammer nose made rectangular; bbl. catch locking slot omitted barrel inscription is in Turkish.	100.00
44 S&W Russian	1870 / 1875	3. Russian-1	6	6½	1-215,704	Same as 3-1	Lanyard ring in butt; sawhandle grip. Made for Russian Army. With Russian inscription. Without Russian inscription	70.00 45.00
44 S&W.R. gallery; 450 Webley	1879 / 1908	New model 3. Single Action	6	4, 5, 6, 6½, 7½, 8	1-38,796	Jan. 17, '65; Jan. 24, '65; July 11, '65; Aug. 24, '69; Apr. 20, '75; Feb. 20, '77; Dec. 18, '77; July 25, '71		45.00
44 Winchester Rifle	1885 / 1908	3.S.A. Frontier	6	4, 5, 6½	1-2,072		Ammo. created need for longer cylinder also long strap bbl.	50.00
38 Winchester Rifle		3.S.A. 38 Winch.	6	6½	75 mfgd.			60.00

Cal.	Year	Model	Cyl.	Barrel	Serial Nos.	Patent Dates	Misc.	Price
32/44 gallery -Target	1887-1910	3.S.A. 32/44	6	6½	2920 mfgd.		299 of these made with long strap, cyl. 1 9/16".	45.00
38/44 gallery -Target	1887-1910	3.S.A. 38/44	6	6½	1413 mfgd.		390 of these made with long strap, cyl. 1 9/16".	45.00
320 S&W Rifle	1880-1887	Rifle 320 Repeating	6	16, 18, 20"	1-977	Jan. 14, 24, '65; July 11, '65; Aug. 24, '69; Apr. 20, '75; Feb. 20, Dec. 18, '77 Reissued July 1871	Designed as 44 S.A. New Model. Detachable stock clamped by thumb screw beneath butt. Only rifle ever made by S.&W. Without stock With stock	55.00 85.00

Cal.	Year	Model	Cyl.	Barrel	Serial Nos.	Misc.	Price
44 S.&W. Russian	1881-1913	44-D.A.-1 Russian	6	4, 5, 6, 6½	1-54,668	275 mfgd. for 38 cal. Winch. rifle with 6½" bbl.	35.00
44 Winch. Rifle	1886-1910	44D.A. Frontier	6	6½	1-15,340	Mfgd. only on long strap frame.	35.00
44 S&W Russian		44-D.A. Wesson Favorite	6		Numbered under same series 44-D.A.1	Bbl. body diameter reduced. Grooves cut in sides of bottom strap. Cyl. diameter reduced.	75.00
44 S&W Spec.	1907	Hand Ejector	6	4, 5, 6½, 7½	13,753 mfgd.	Also known as 'Triple Lock.' 5,000 were mfgd. for British Army. In 1915 the model was changed to correct close condition of cartridge head in cyl. and 73,650 mfgd.	
450 Eley		New Century	6		1,226		
45 Colt			6		21		
455 Mark II			6		5,000		50.00
44 S&W Spec. 44 S&W Russian 45 Colt		Hand Ejector Second Model	6	4, 5, 6	From No. 15,525	727 mfgd. in Cal. 45 Colt with 6½" bbl. Designed as British Army Model except hammer shaped to engage trigger notch for increased D.A. throw.	50.00

APPENDIX I
1876 ORDNANCE TEST

Dimensions:	Schofield	Colt	Remington
Total length, inches	12.5	12.5	13.03
Length of barrel, inches	7.	7.5	7.46
Diameter of bore, inches	0.435	0.45	0.44
Grooves, depth, inches	0.0075	0.005	0.004
Grooves, number of	5	6	5
Twist	Right	Left	Left
Uniform, one turn in inches	20	16	
At base, one turn in inches			5
At muzzle, one turn in inches			26
Weights:			
Total weight, pounds	2.5	2.31	2.6
Weight of powder charge, gr.	28	28	28.3
Weight of bullet, grains	230	230	226.6

Firing Tests:	Initial Velocity, ft.	Penetration	Recoil
Schofield, service ammunition	744	46	57
Colt, service ammunition	733	47	56
Colt, Bridgeport ammunition	890	58	74
Remington, Remington ammunition	655	43	53

Bridgeport ammunition is 37.6 grs. powder and 249.6 grs. lead

APPENDIX II

The various stampings to be found on Government issue "Schofield Smith & Wesson."

1. "U.S." on butt of frame.
2. Proof letter "P":
   a. On barrel underside projection, right side;
   b. On cylinder 0.1" from rear in center line of a groove.
3. Subinspector's initials:
   a. On frame, right side;
   b. On barrel just forward of proofmarks;
   c. On cylinder 0.1" from rear in center line of a groove.
   d. On stock 0.1" from butt, right side (also on left as receiving stamp if subinspector receives arm).
4. If inspected by an officer, his initials as receiving stamp on stock 0.1" from butt, left side, and year of inspection above such initials.
5. Letter "c" stamped on all condemned parts.

APPENDIX III
BALLISTICS OF SMITH & WESSON REVOLVER CARTRIDGES

Cartridges	Bullet	Bullet Grains	Muzzle ft. sec.	Muzzle ft. pds.	Range yard	Barrel inches
.22 Short	Lead	29	865	48	50	6
.22 Short Hi-Speed	Lead	29	1035	69	75	6
.22 Long Hi-Speed	Lead	29	1095	77	50	6
.22 Long Rifle	Lead	40	970	83	200	6
.22 Long Rifle Hi-Speed	Lead	40	1125	112	300	6

.32 S.&W.	Lead	88	720	98	75	3
.32 S.&W. Long	Lead	98	795	138	125	4
.32 S.&W. Long Wad Cutter	Lead	98	770	129	125	4
.38 S.&W.	Lead	146	745	179	100	4
.38 S.&W. Special	Lead	158	870	266	300	6
.38 S.&W. Special	Metal Point	158	870	266	300	6
.38 S.&W. Special	Lead	200	745	247	300	6
.38 S.&W. Mid-Range Wad Cutter	Lead	148	770	193	100	6
.38 S.&W. Special Hi-Speed	Lead	158	1115	436	500	5
.38 S.&W. Special Hi-Speed	Metal Point	150	1115	436	500	5
.44 S.&W. Russian	Lead	246	770	324	225	6½
.44 S.&W. Special	Lead	246	770	324	225	6½
.45 Auto. Rim	Lead	230	820	343	100	5½
.45 A.C.P.	Metal Point	230	860	378	100	5½
.45 Auto. Wad Cutter	Metal Point	185	870	311	100	5½
.357 Magnum	Lead	158	1450	690	600	8⅜
.357 Magnum	Metal Point	158	1450	690	600	8⅜

APPENDIX IV

Points of Sure-Fire Precision Peculiar to Smith & Wesson Revolvers as set forth in Smith & Wesson Centennial Catalog:

1. FRONT SIGHT. On military and police type arms, the front sight is an integral part of the barrel forging, and is therefore immovable. On target arms the front sight blade and the sight base are a single unit, and this unit is firmly pinned into the barrel rib which is an integral part of a target barrel forging. This insures rugged construction and eliminates the hazard of lost or loosened sights.

2. UNIQUE FORWARD CYLINDER LOCK. The only American revolver in which the cylinder is locked in perfect alignment with barrel by two-point suspension operating in hardened steel bearings, which also insures protection of extractor rod.

3. RIGHT HAND BARREL THREAD AND RIFLING. Combination of barrel threading and rifling plus torsional force of bullet maintains the firm seating of barrel shoulder against the frame.

4. HEAT-TREATED MODERN ALLOYED STEEL CYLINDER. Cylinders are built to withstand continual shock. They are made of the very finest shock resistant steel obtainable.

5. LIGHTNING FAST SPRING-OPERATED CYLINDER STOP. This lockwork allows for lightning-like rapid fire; functions perfectly at highest firing rate ever obtained with any handgun.

6. RATCHET, EXTRACTOR HEAD AND STEM. Permanently joined to withstand roughest treatment; no malfunctioning, no misalignment, after years of service.

7. ACCURATE REAR CYLINDER LOCK. Perfect alignment of cylinder and barrel assured by mounting front and rear cylinder locks in hardened steel bearings.

8. THUMB PIECE (ATTACHED TO BOLT) OPERATED THROUGH FRAME. Always tight, always safe. S.&W. thumb piece operates through the frame—no loose pieces when the sideplate is removed.

9. HAMMER AND TRIGGER GLASS-HARD. The extreme hardness of hammer notches and trigger points, made of case-hardened low carbon manganese steel, assures a crisp, unchanging trigger pull, even after years of service.

10. SIMPLE, POSITIVE HAMMER BLOCK. No fragile stampings, no complicated link motions, no dependence on frail pins. S.&W. hammer blocks are strong, simple, sure.

11. SMOOTHEST DOUBLE ACTION. Balanced spring tensions produce smooth uniform double action at all firing speeds.

12. MAIN SPRING ANCHORAGE. Solidly anchored at the foot, the main spring is positively adjusted at the factory to proper tension.

13. TRIGGER SPRING AND HAMMER REBOUND. Rugged, positive construction permits fastest double action. (5 shots and 5 hits in less than ½ second.)

14. IMPORTANT WORKING POINTS. Assuring long life and perfect functioning, S.&W. working points are full width of the lockwork. There are no knife-edged contacts or minute working parts.

APPENDIX V

Year Introduced—Caliber—Name of Model—Changes in Model

1857—22—1-1: 1-2: 1-3.
1861—32—2-OLD
1865—32—1½-OLD: 1½-NEW MODEL.
1869—44—3-1 AMERICAN. 3-2.
1870—44—S.A. RUSSIAN: TURKISH
1875—45—SCHOFIELD S.A.
1876—38—2-1 S.A. 2-2 2-3:1891.
1877—38—38 D.A.-1 38 D.A.-2; 3; 4; 5.
1878—44—44 S.A. NEW MODEL.
1880—320—Repeating Rifle
1880—32—32 D.A.-1 32 D.A.-2: D.A.-3.
1881—44—44 D.A.-1
1885—44 WINCHESTER—44 S.A. FRONTIER
1885—38 WINCHESTER—44 S.A. NEW MODEL
1886—44 WINCHESTER—44 D.A. FRONTIER: WESSON FAVORITE
1887—38/44—Model 3-38/44—S.A.
1887—32/44—44 S.A. NEW MODEL
1887—38—38 SAFETY-1. -2; -3; -4; 5; (also known as New Departure, Hammerless)
1888—32—32 SAFETY-1. -2.
1891—38—1891-3 MODEL MEXICAN
1893—22—22 S.S.1. S.S.2; S.S. PERFECTED TARGET PISTOL.
1896—32—M I-H.E. MODEL 1903. 1903-1; -2; -3; -4; -5; -6.
1896—32—32 I, HAND EJECTOR 1
1899—32/20-32/20 Hand Ejector Winc. 1; 2-1902; 1903; 1905; 1905-1; -2; -3; -4.
1899—38—38 HAND EJECTOR, Military & Police 1, 1902; 1903; 1905; -1; -2; -3; -4.
1902—22—M-HAND EJECTOR 1.; -2; 3(PERFECTED HAND EJECTOR).
1907—44—H.E. NEW CENTURY H.E.-2nd MODEL.
1909—38—38 D.A. PERFECTED
1911—22/32-H.E. TARGET
1911—22—22/32 HAND EJECTOR (HEAVY DUTY FRAME)
1913—35—AUTOMATIC
1915—455—HAND EJECTOR. MARK 2. ENGLISH.
1917—32—I. H. E. 3 (REGULATION POLICE)
1917—45—H.E.-U.S. REVOLVER

1924—32—AUTOMATIC
1925—22—22 STRAIGHT LINE
1926—44—44 TARGET
1930—38/44-38/44 OUTDOORSMAN
1931—22—K-22 OUTDOORSMAN
1933—38—38 MILITARY & POLICE (CALLED "M & P" IN 1899)
1935—357—MAGNUM
1936—38/32-38/32 TERRIER
1937—22/32-22/32 KIT GUN
1938—38—38 MILITARY & POLICE (STILL IN PRODUCTION)
1940—22—K-22 MASTERPIECE
1940—38—K-200
1942—38—VICTORY
1950—45—ARMY
1950—38—38 CHIEFS SPECIAL
1952—38—38 CENTENNIAL

Copyrighted 1953.

THE SELF-MADE MEN OF OUR TIMES.—Horace Smith, Springfield, Massachusetts.

Daniel B. Wesson

DANIEL BAIRD WESSON

CARL R. HELLSTROM
President, Smith & Wesson

NEW SMITH & WESSON FACTORY—SPRINGFIELD, MASS.

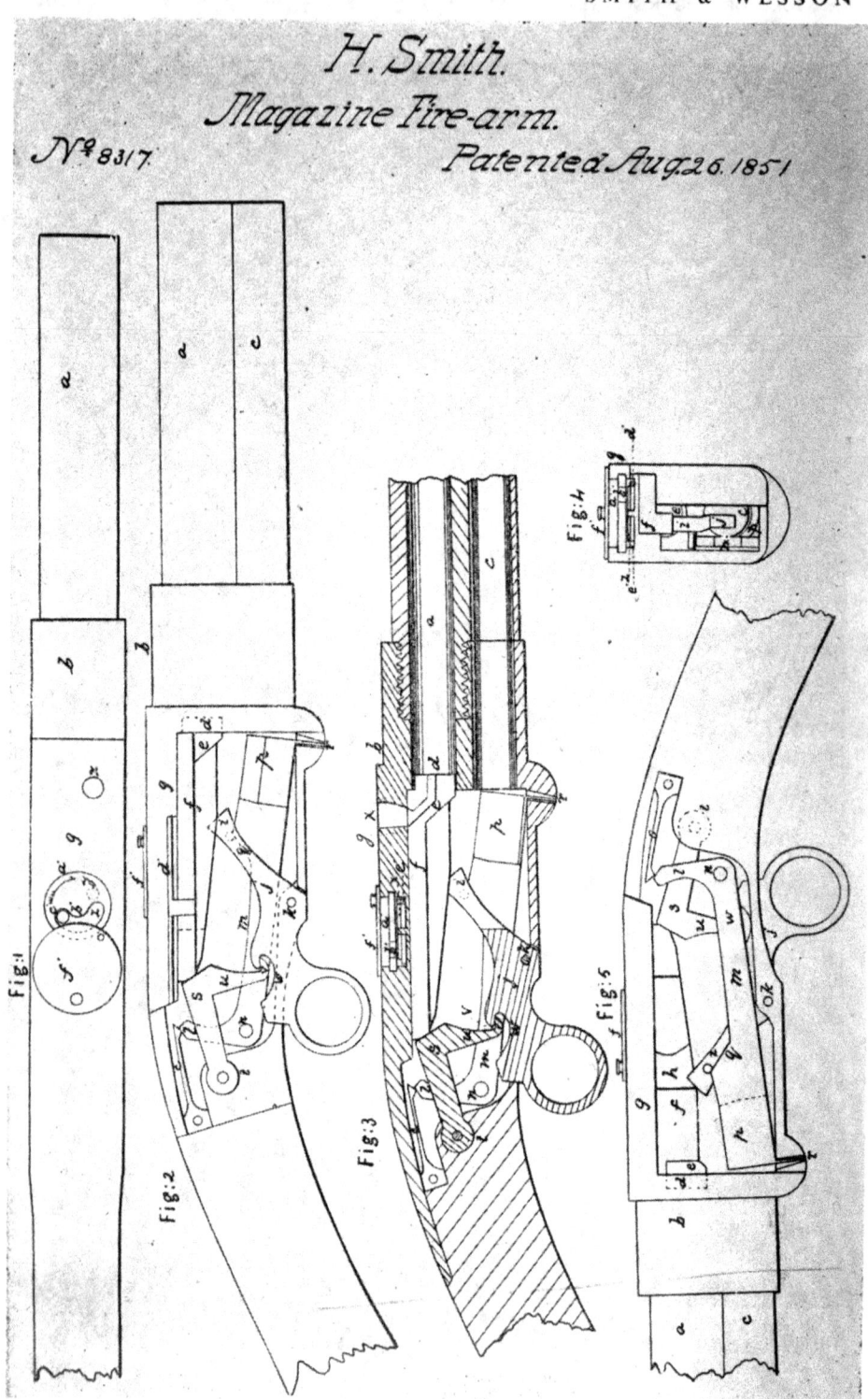

H. SMITH & D. B. WESSON
MAGAZINE FIREARM.

No. 10,535. Patented Feb. 14, 1854.

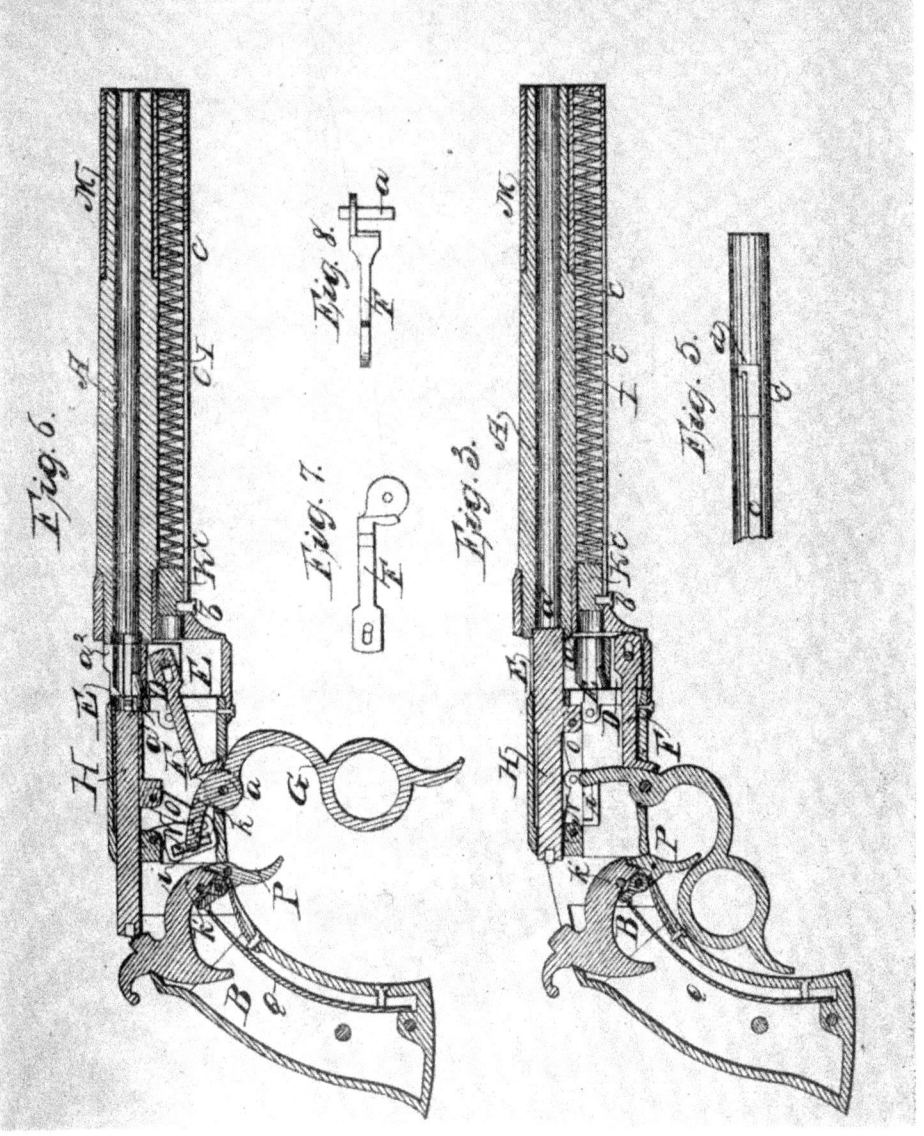

VOLCANIC MAGAZINE PISTOL MARKED "SMITH & WESSON, PATENT, NORWICH, CONN."
—Courtesy Smithsonian Institution

THE STORY OF THE REVOLVER 105

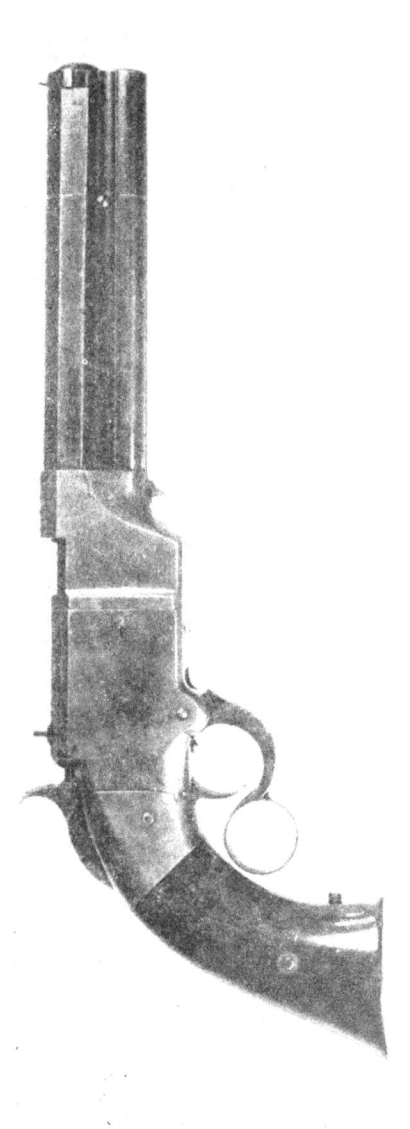
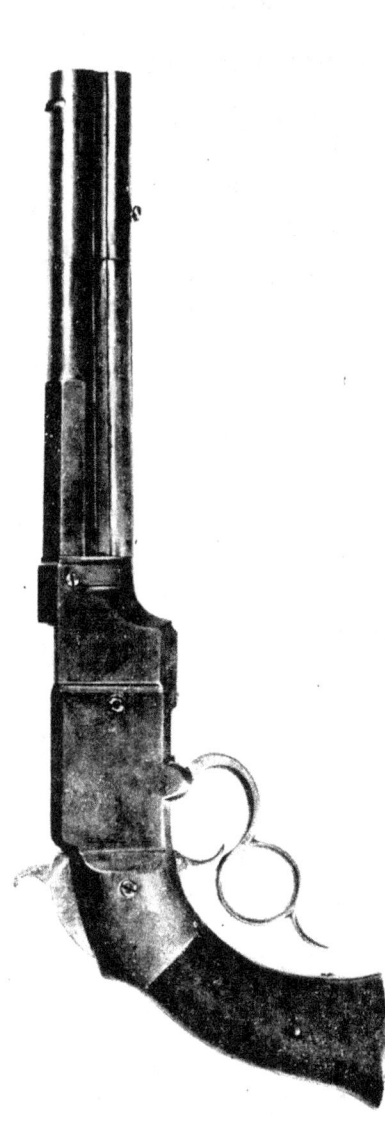

Bottom: SMITH & WESSON—CAL. 30
—Courtesy Smithsonian Institution

Top: VOLCANIC—CAL. 36

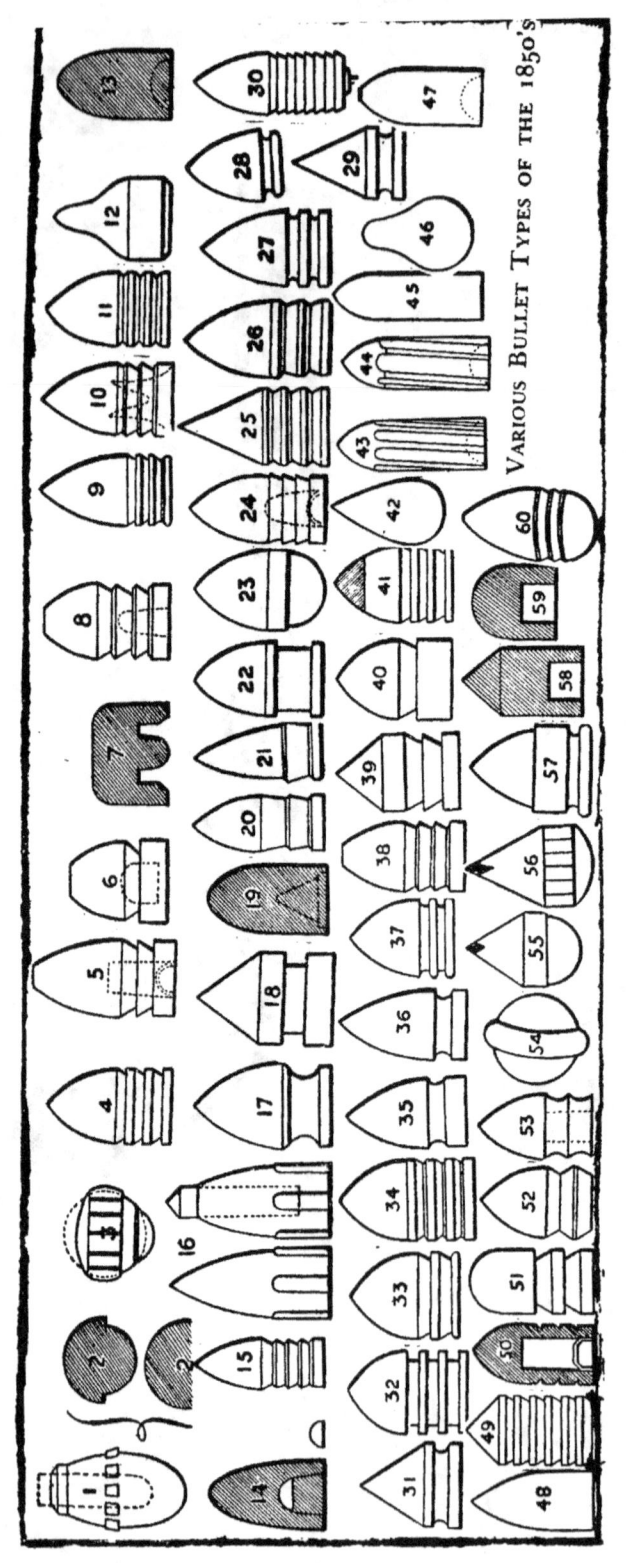

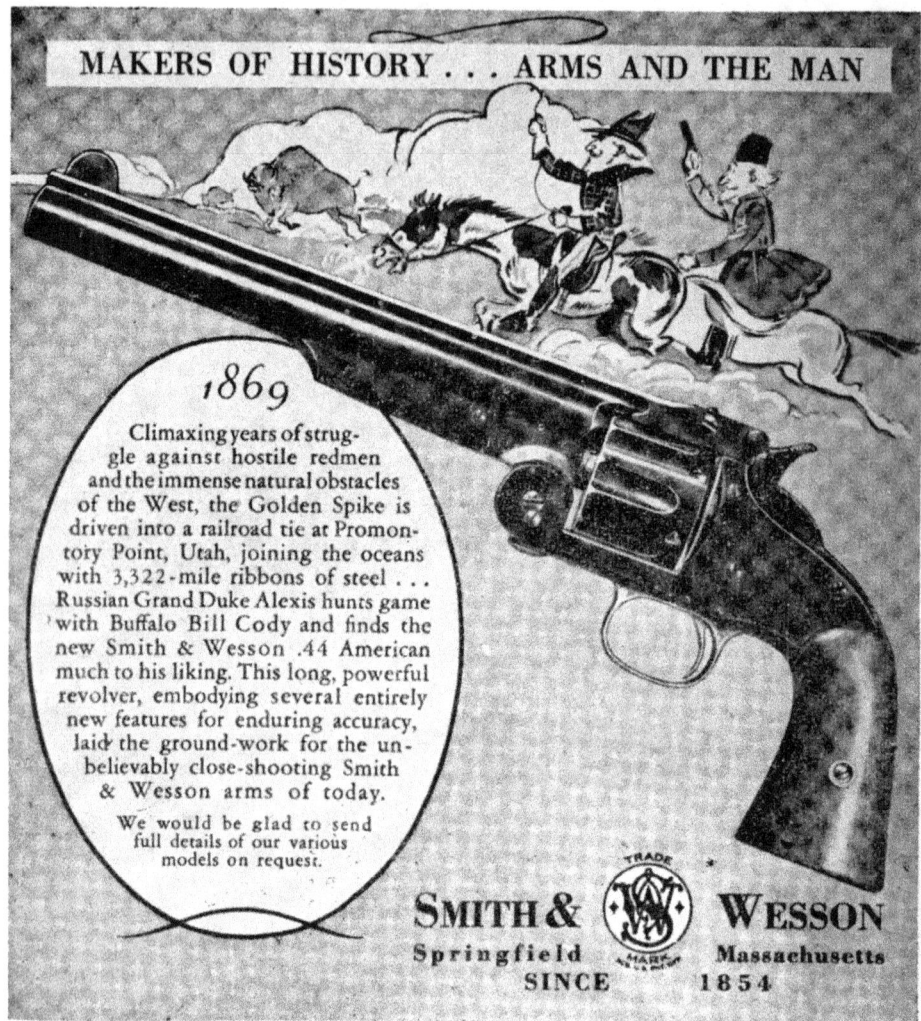

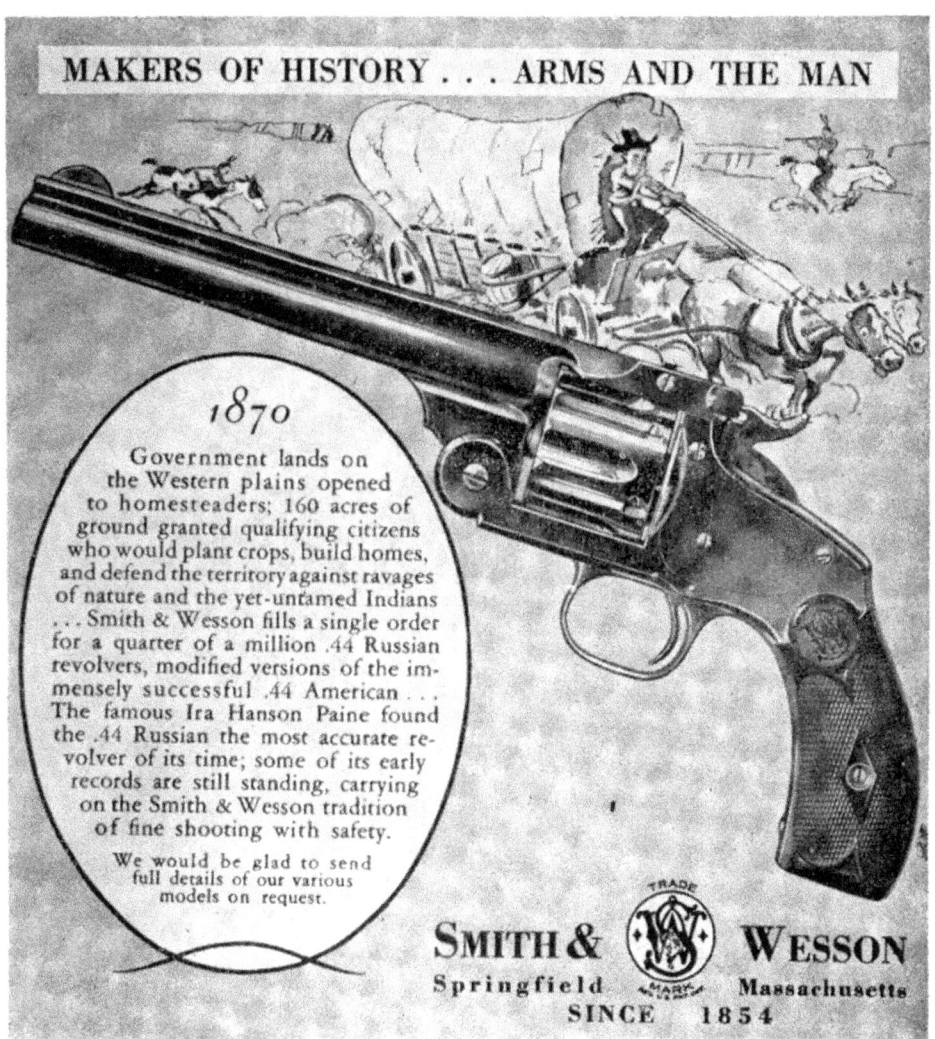

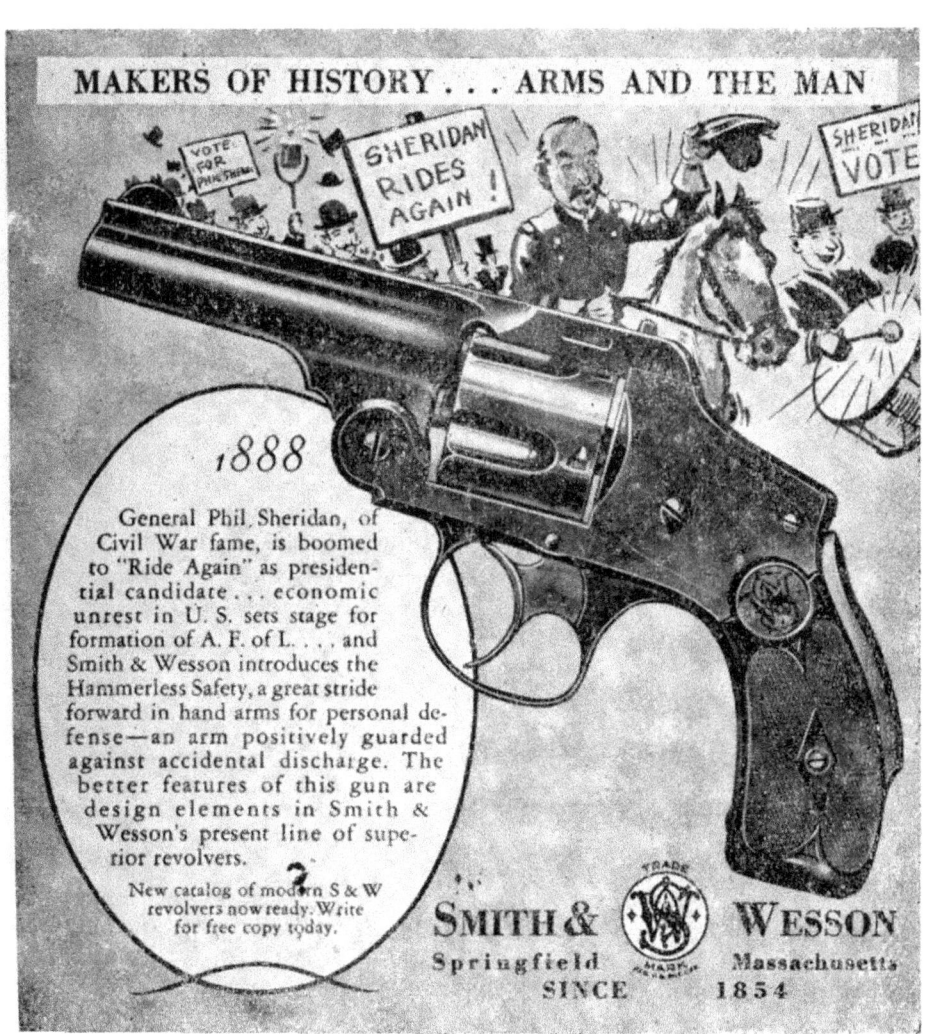

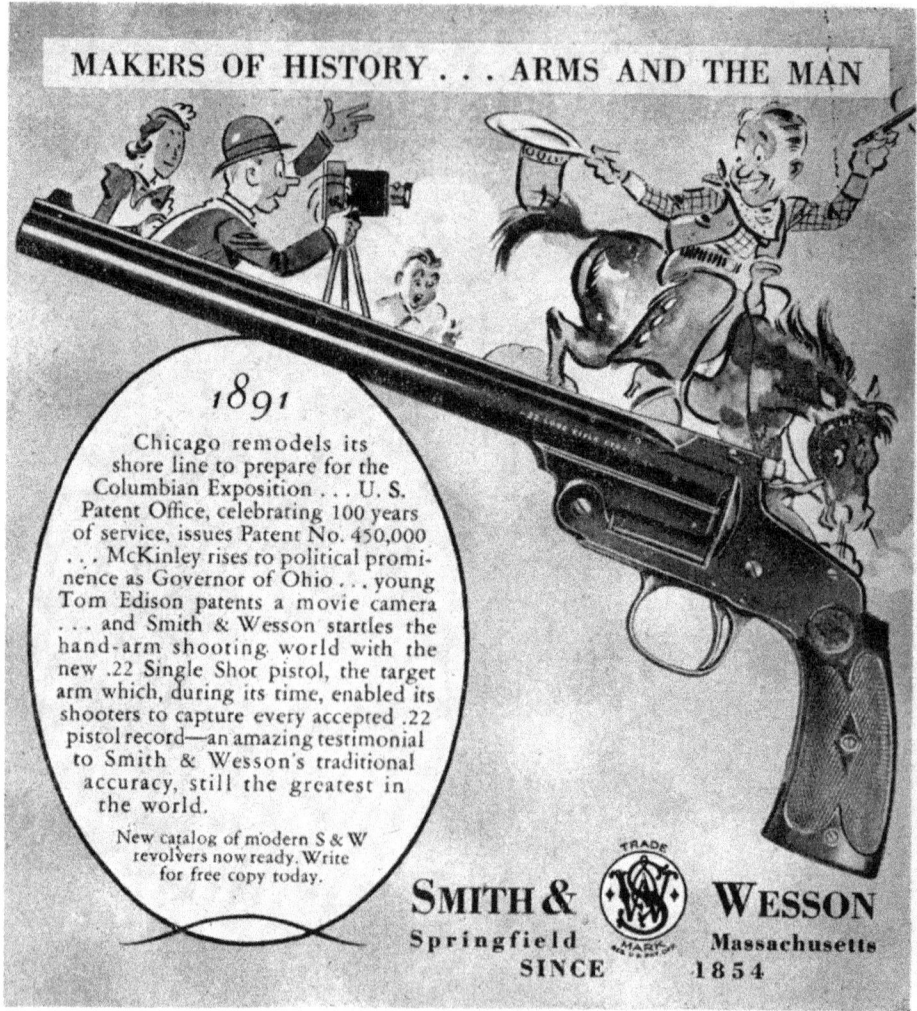

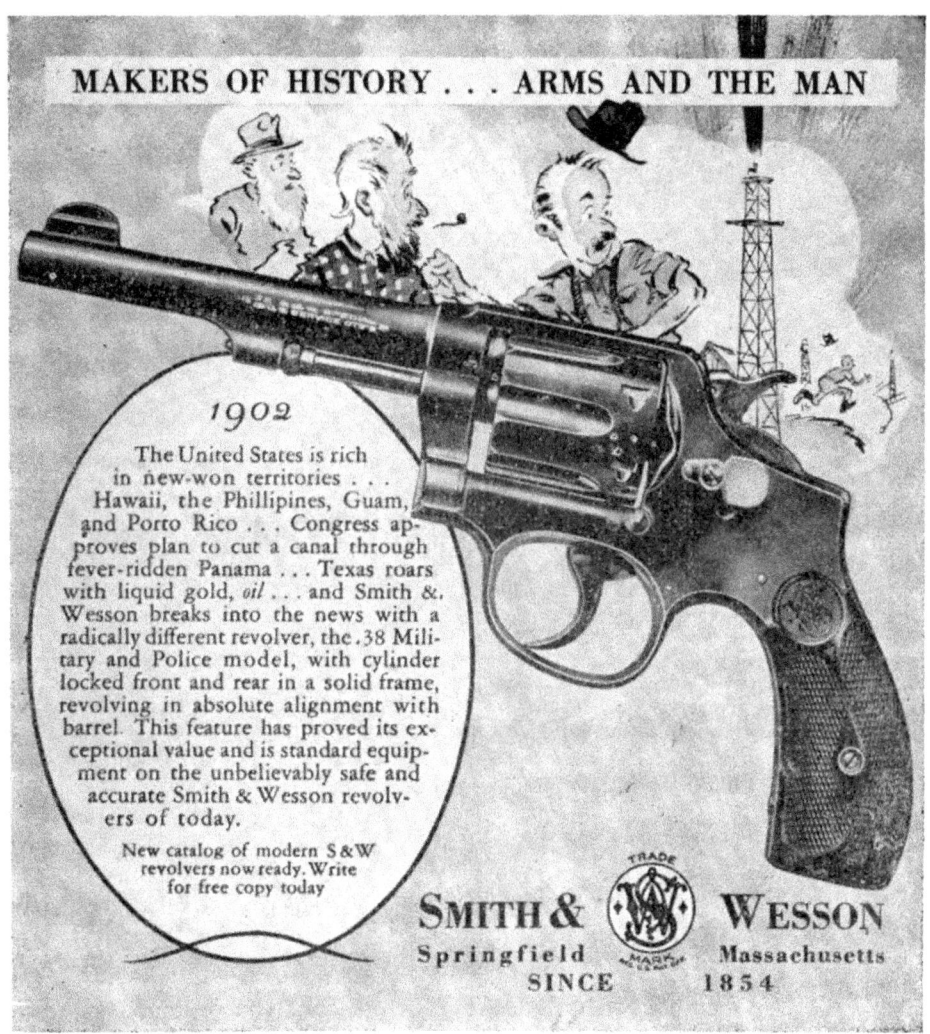

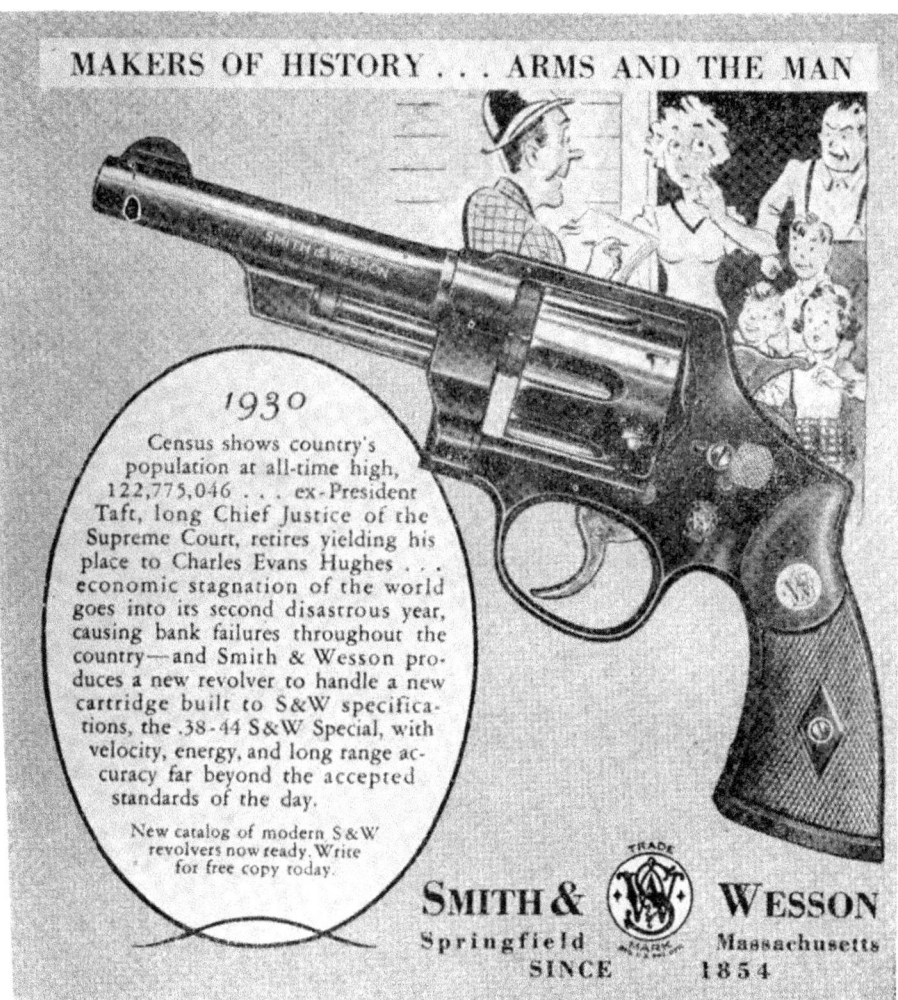

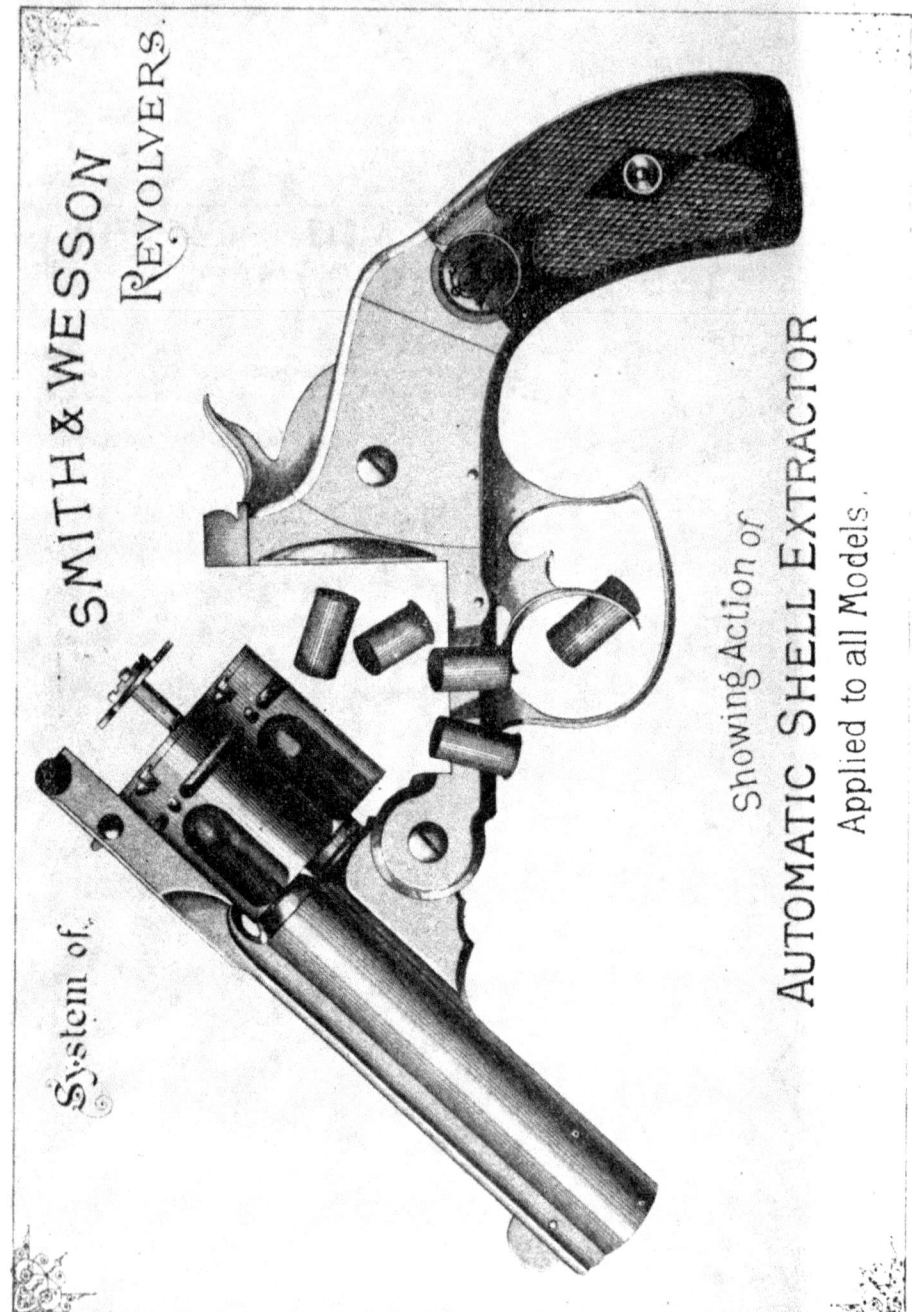

THE STORY OF THE REVOLVER 121

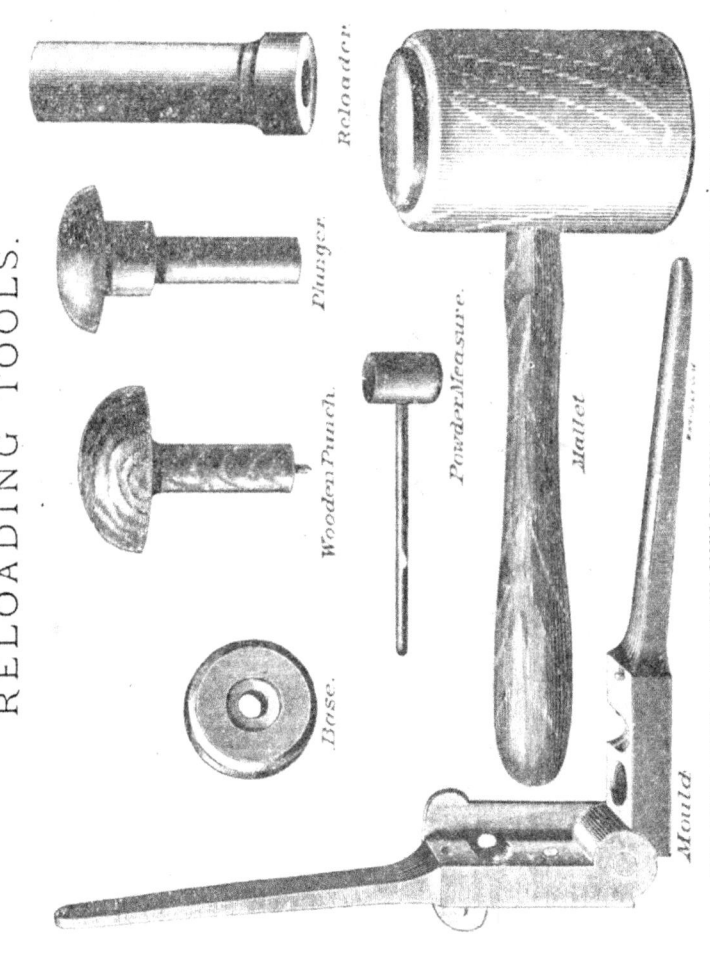

RELOADING TOOLS.

FOR RELOADING CARTRIDGE SHELLS USED IN SMITH & WESSON REVOLVERS.

These Cartridge Shells can be Reloaded a Number of Times with Perfect Safety, making a Great Saving in Cost of Ammunition.

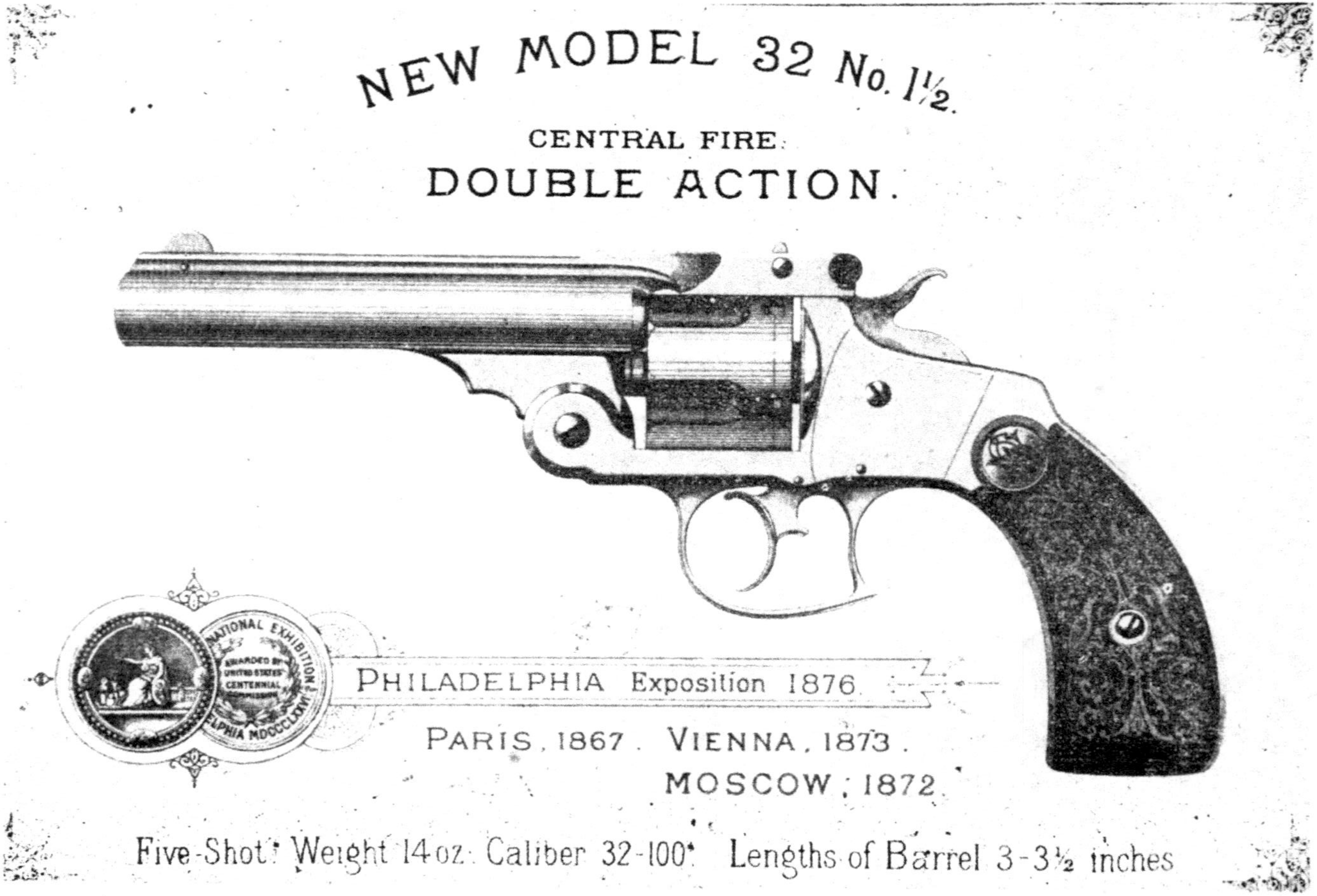

SMITH & WESSON

NEW MODEL 32 No. 1½.
CENTRAL FIRE.
DOUBLE ACTION.

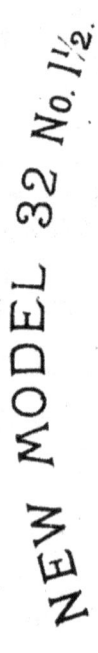

PHILADELPHIA Exposition 1876. PARIS, 1867. VIENNA, 1873. MOSCOW, 1872.

Five-Shot. Weight 14 oz. Caliber 32-100". Lengths of Barrel 3-3½ inches

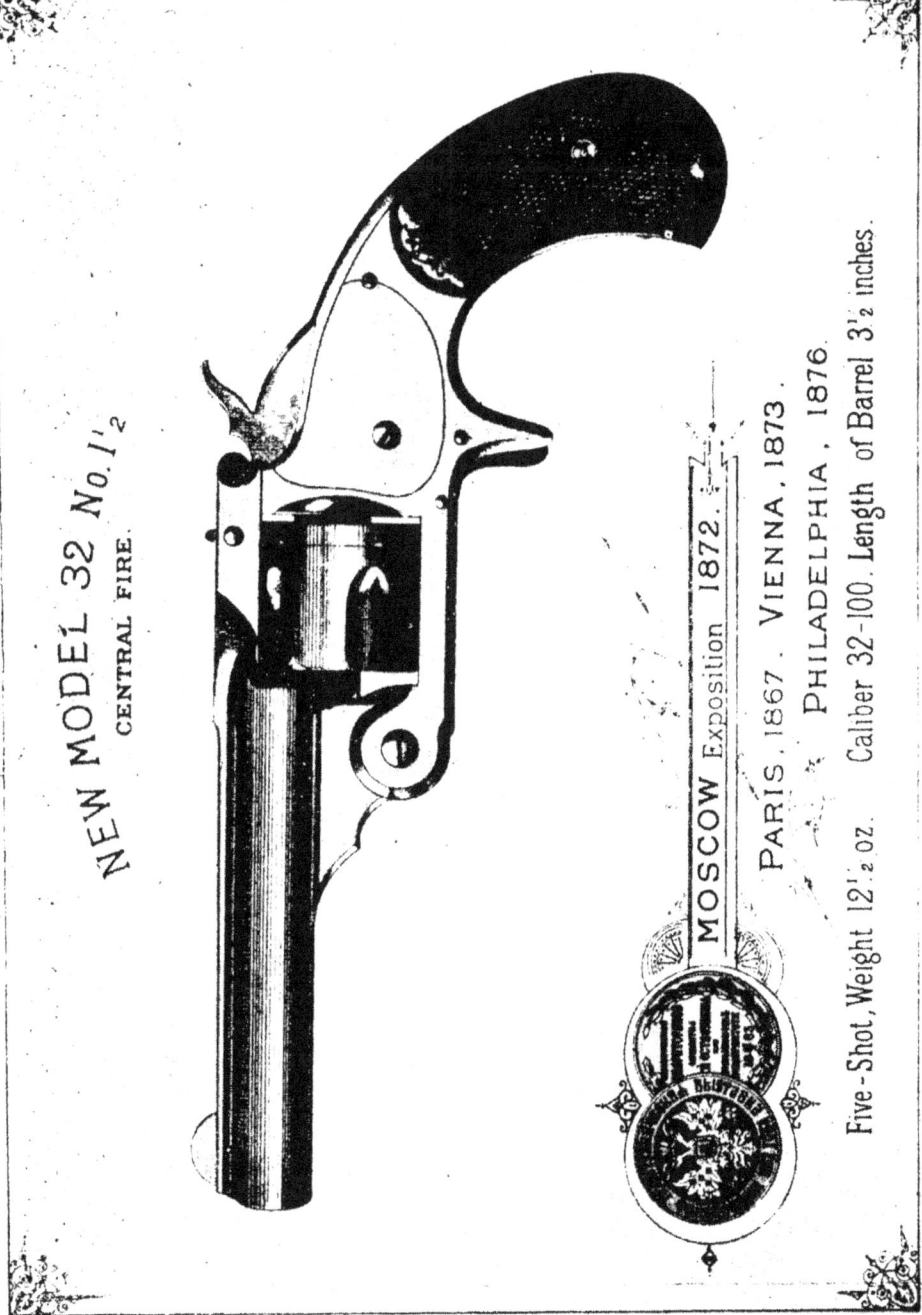

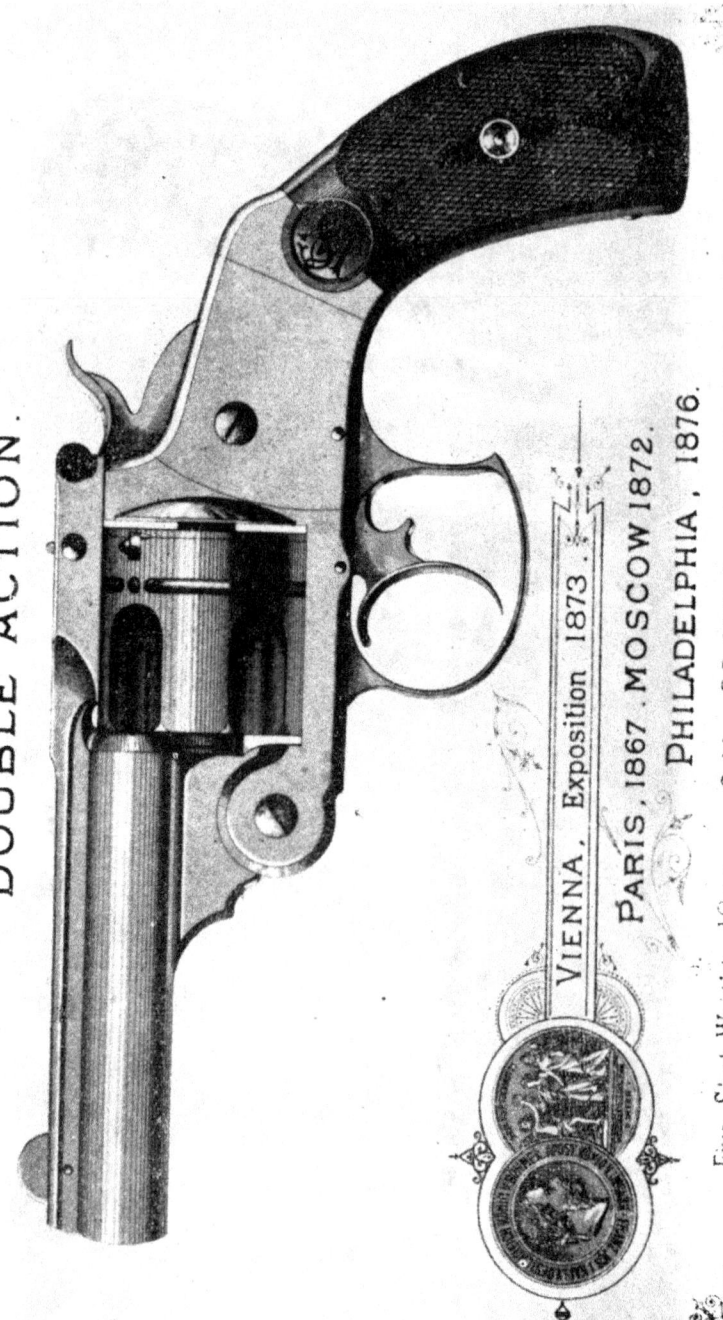

NEW MODEL 38 No. 2.
CENTRAL FIRE.
DOUBLE ACTION.

VIENNA, Exposition 1873.
PARIS, 1867. MOSCOW 1872.
PHILADELPHIA, 1876.
Caliber 38-100. Lengths of Barrel 3¼-4-5 inches.
Five-Shot. Weight 18 oz.

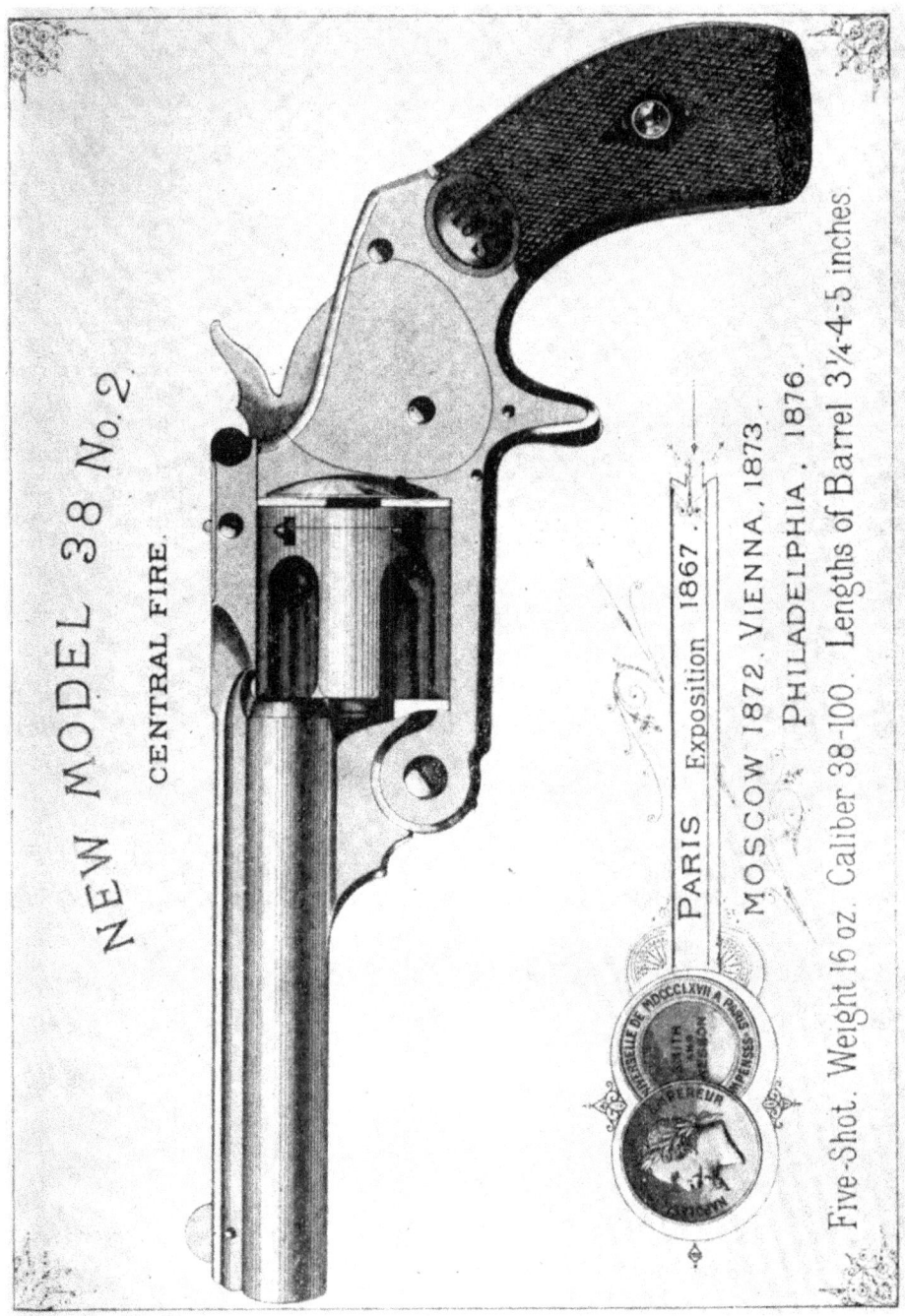

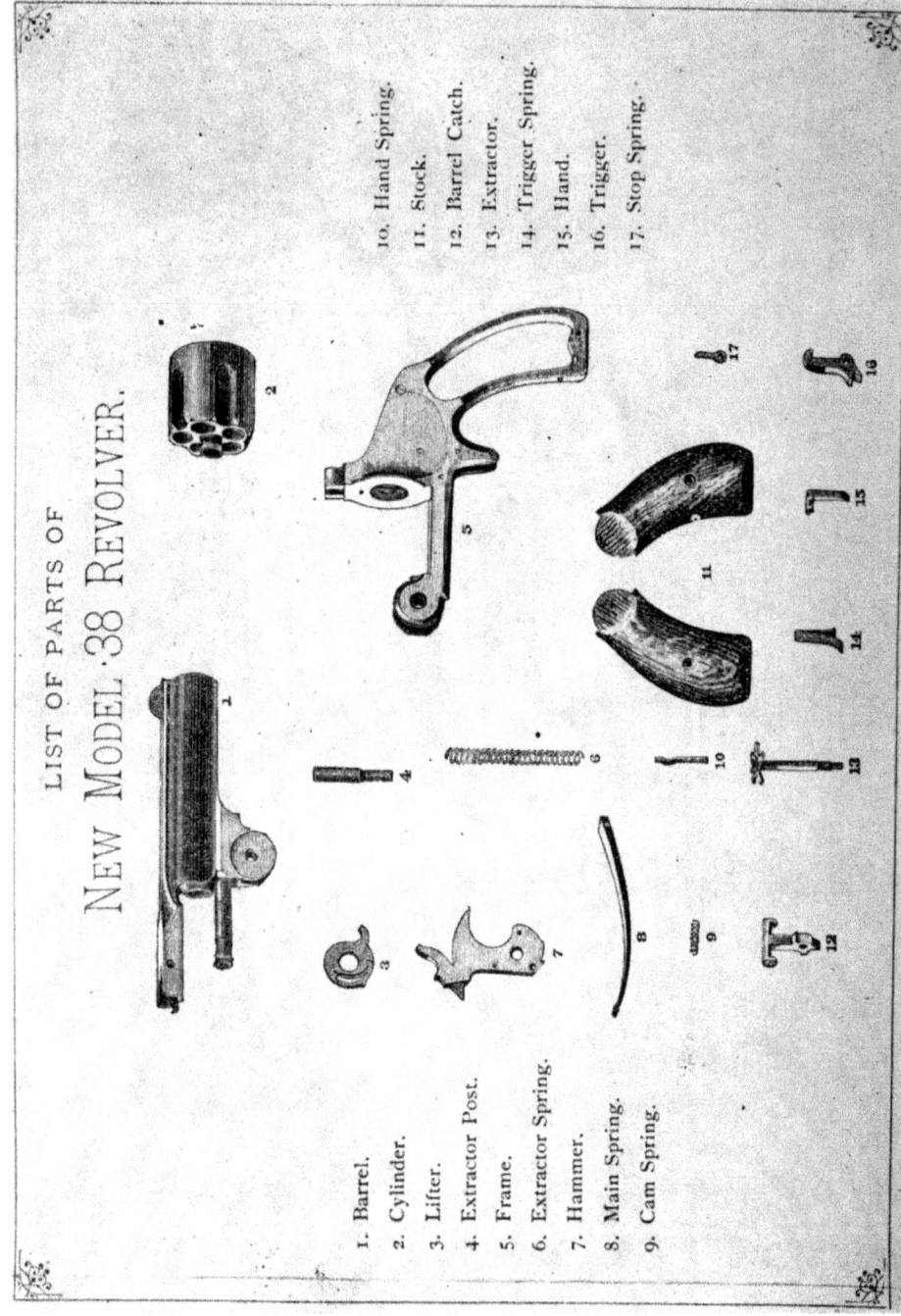

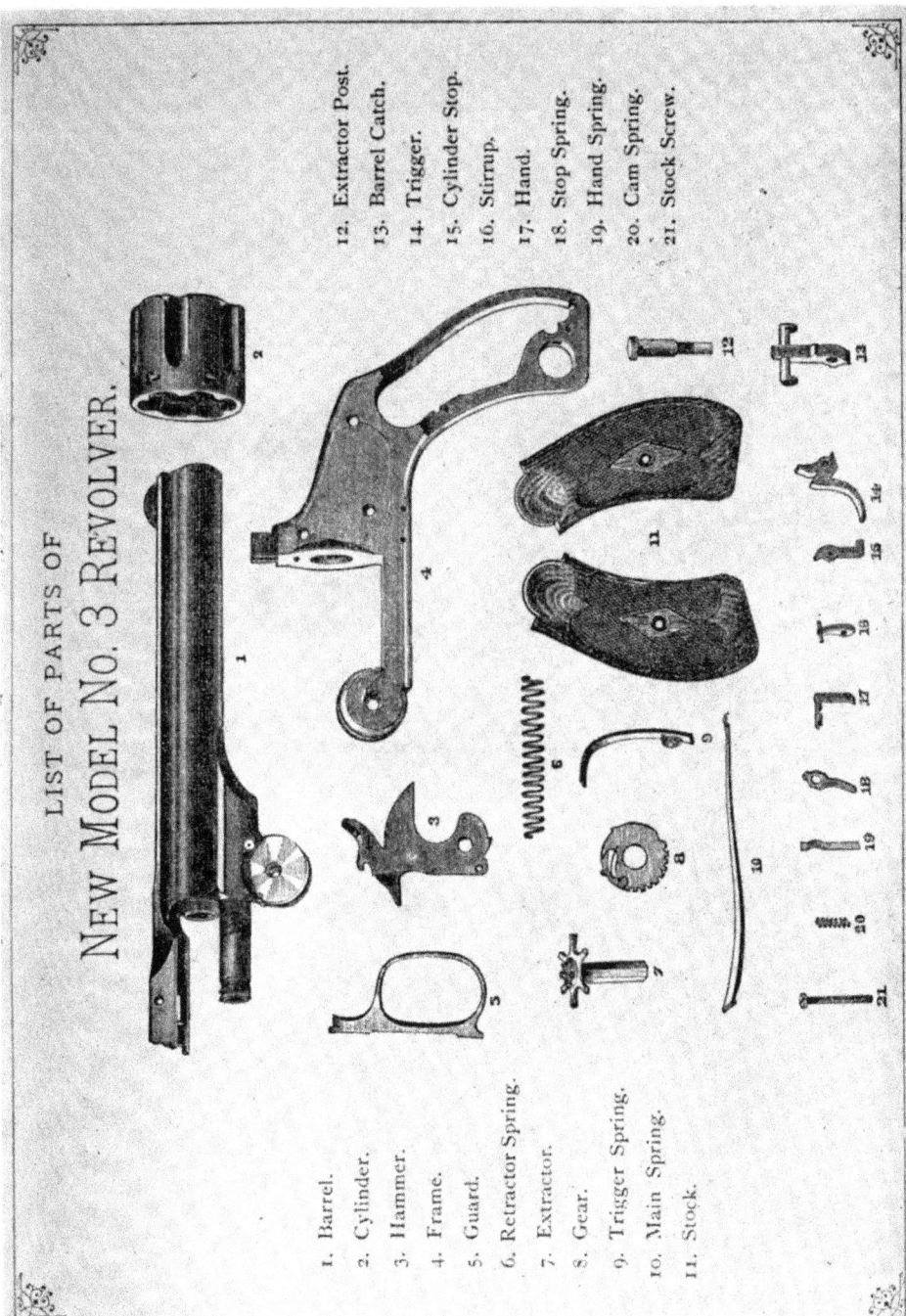

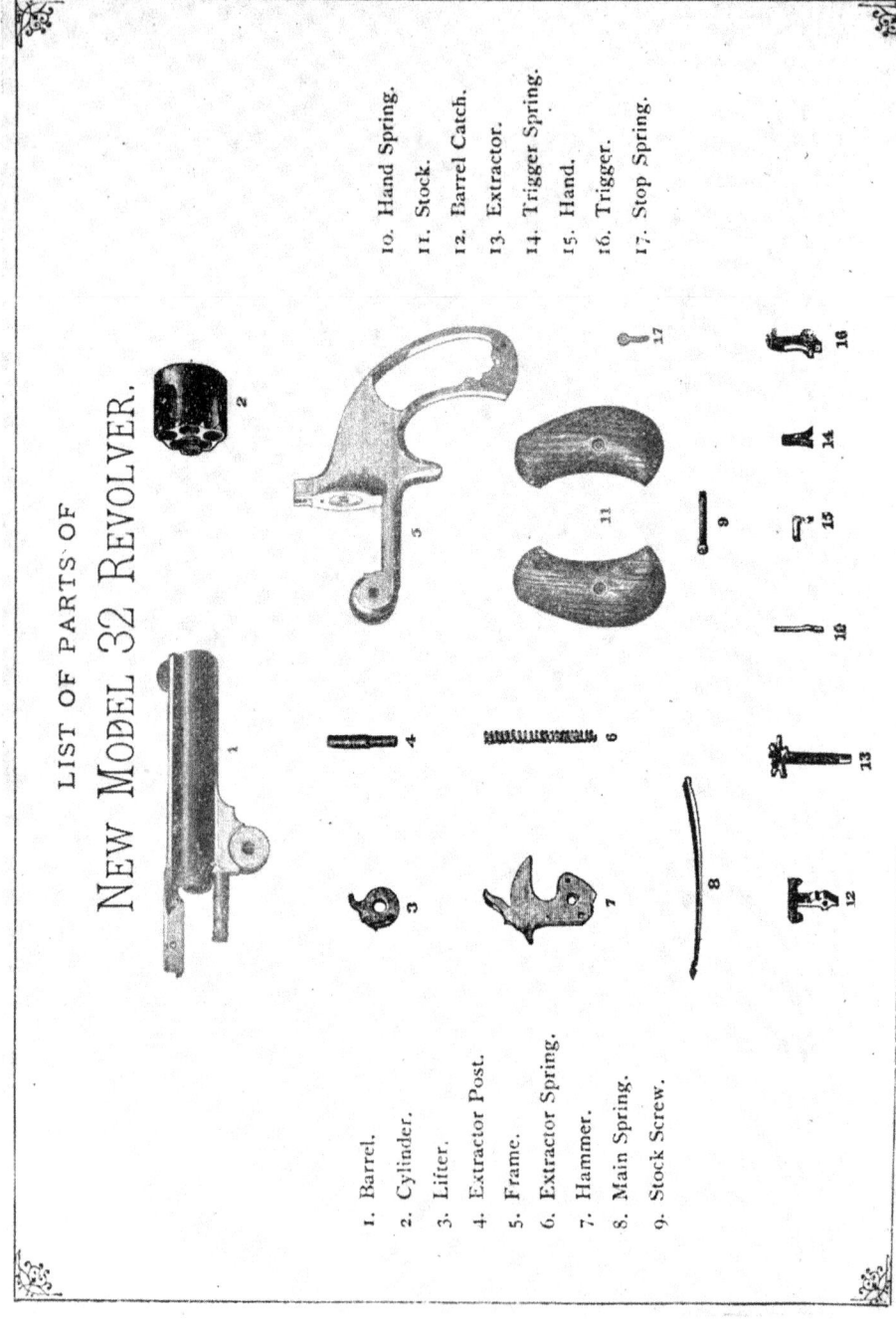

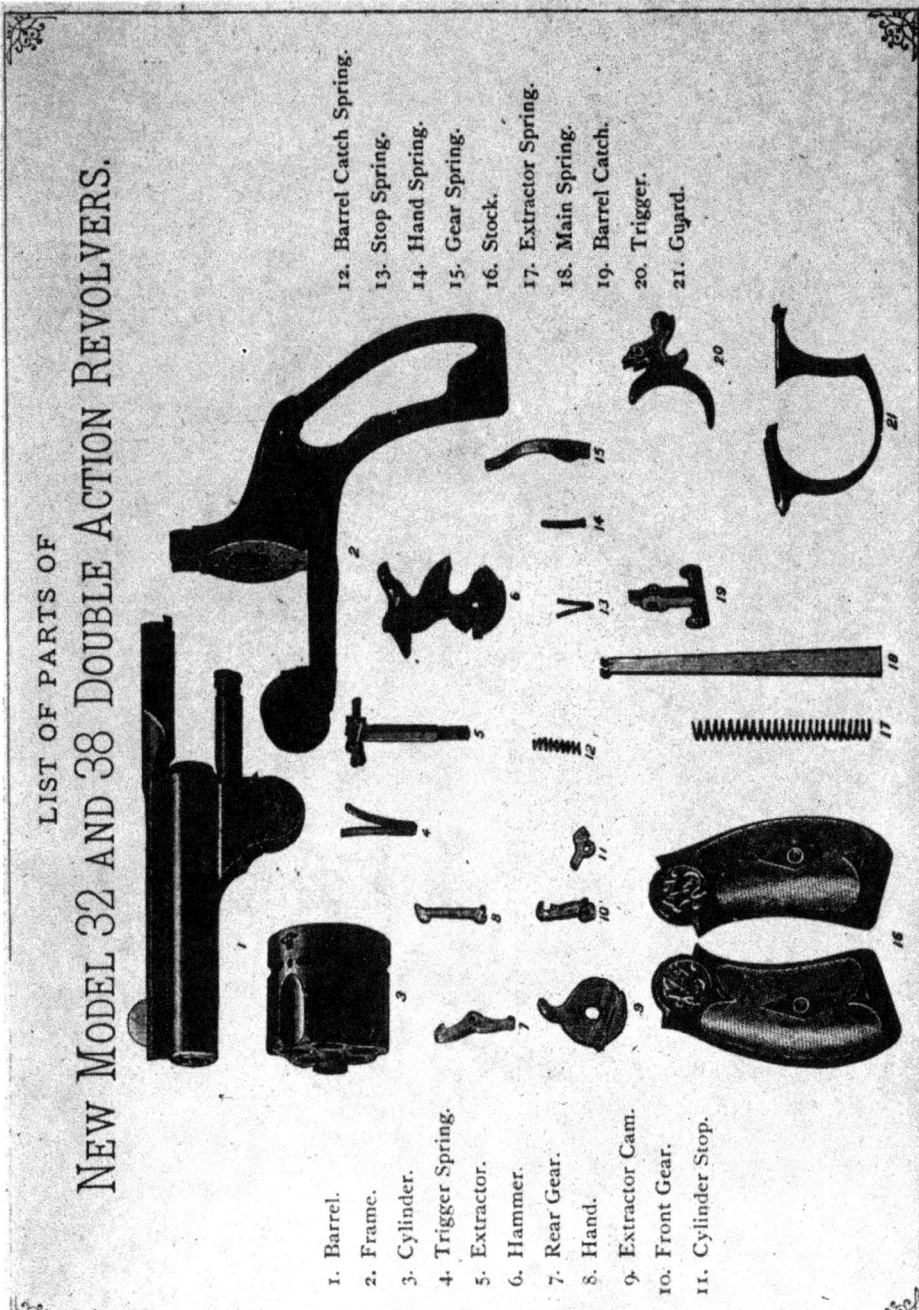

LIST OF PARTS OF
NEW MODEL 32 AND 38 DOUBLE ACTION REVOLVERS.

1. Barrel.
2. Frame.
3. Cylinder.
4. Trigger Spring.
5. Extractor.
6. Hammer.
7. Rear Gear.
8. Hand.
9. Extractor Cam.
10. Front Gear.
11. Cylinder Stop.
12. Barrel Catch Spring.
13. Stop Spring.
14. Hand Spring.
15. Gear Spring.
16. Stock.
17. Extractor Spring.
18. Main Spring.
19. Barrel Catch.
20. Trigger.
21. Guard.

No. 1, Pocket Pistol.

Seven-Shot, Weight 8 oz., Wrought Iron Frame, Rosewood Stock, Caliber 22-100, Length of Barrel 3 Inches.

LIST OF PARTS

1. Barrel.
2. Frame.
3. Cylinder.
4. Side Plate.
5. Hammer.
6. Trigger.
7. Rammer Pin.
8. Cylinder Stop.
9. Barrel Catch.
10. Hand.
11. Stirrup.
12. Main Spring.
13. Barrel Catch Spring.
14. Cylinder Stop Spring.
15. Hand Spring.
16. Trigger Spring.
17. Cam Spring.
18. Hammer Stud.
19. Trigger Pin.
20. Stop Pin.
21. Stirrup Pin.
22. Hand Spring Pin.
23. Stock Screw.
24. Barrel Joint Screw.
25. Rammer Pin Screw.
26. Plate Screw.
27. Strain Screw.
28. Cylinder Screw.
29. Cam Spring Screw.
30. Barrel Catch Screw.
31. Escutcheon Nut.
32. Escutcheon.
33. R. H. Stock.
34. L. H. Stock.

EXTENSION STOCK.

Adapted to No. 3, or Army Revolver.

CAUTION.

Cartridges for SMITH & WESSON Revolvers, (when properly made,) are not CRIMPED, but made straight, as shown in sketch No. 1. It is the practice of some manufactures of these cartridges to crimp them as shown in sketch No. 2. Such cartridges are of an inferior quality. They will not make accurate shooting, nor CAN THEY BE RELOADED.

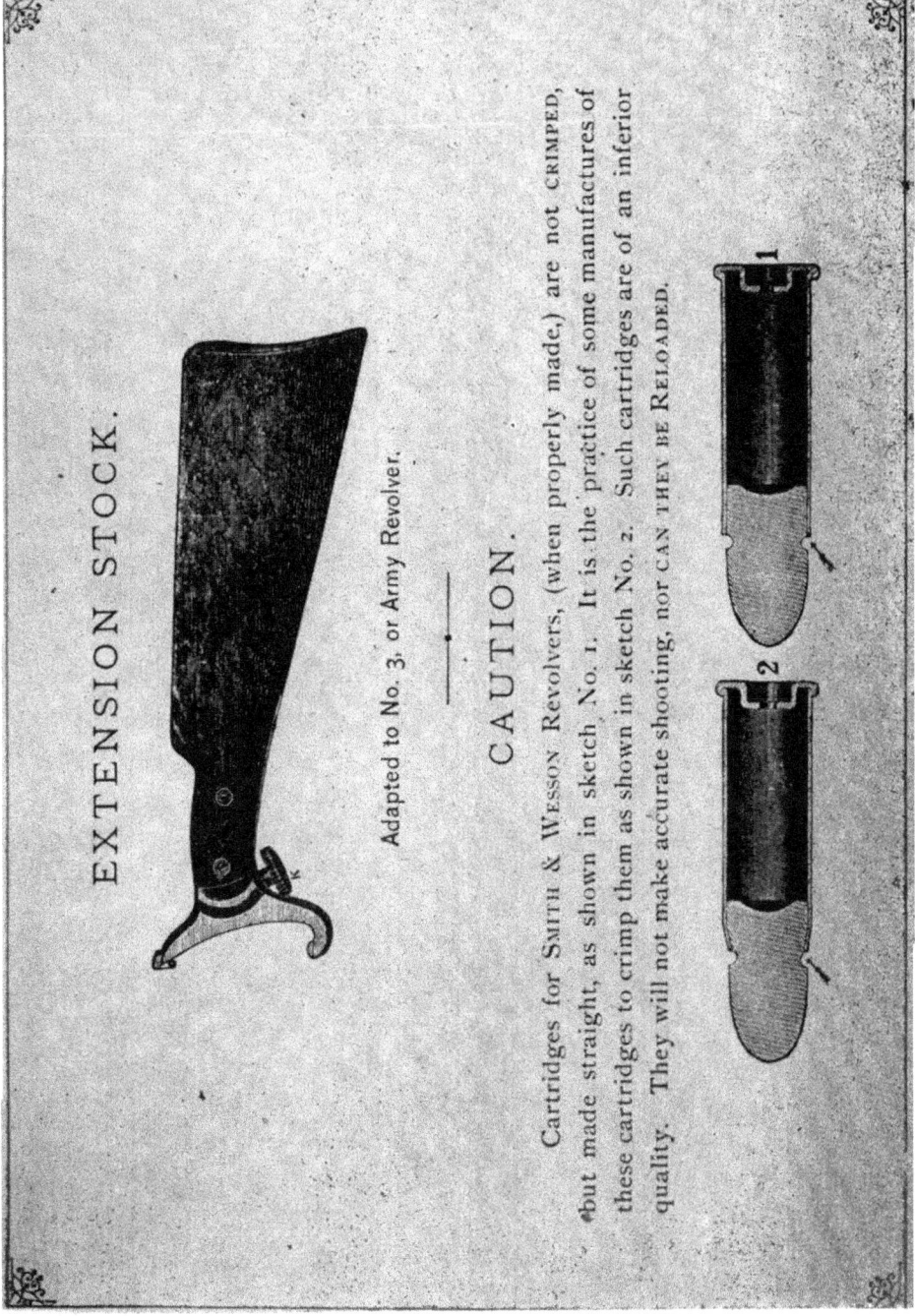

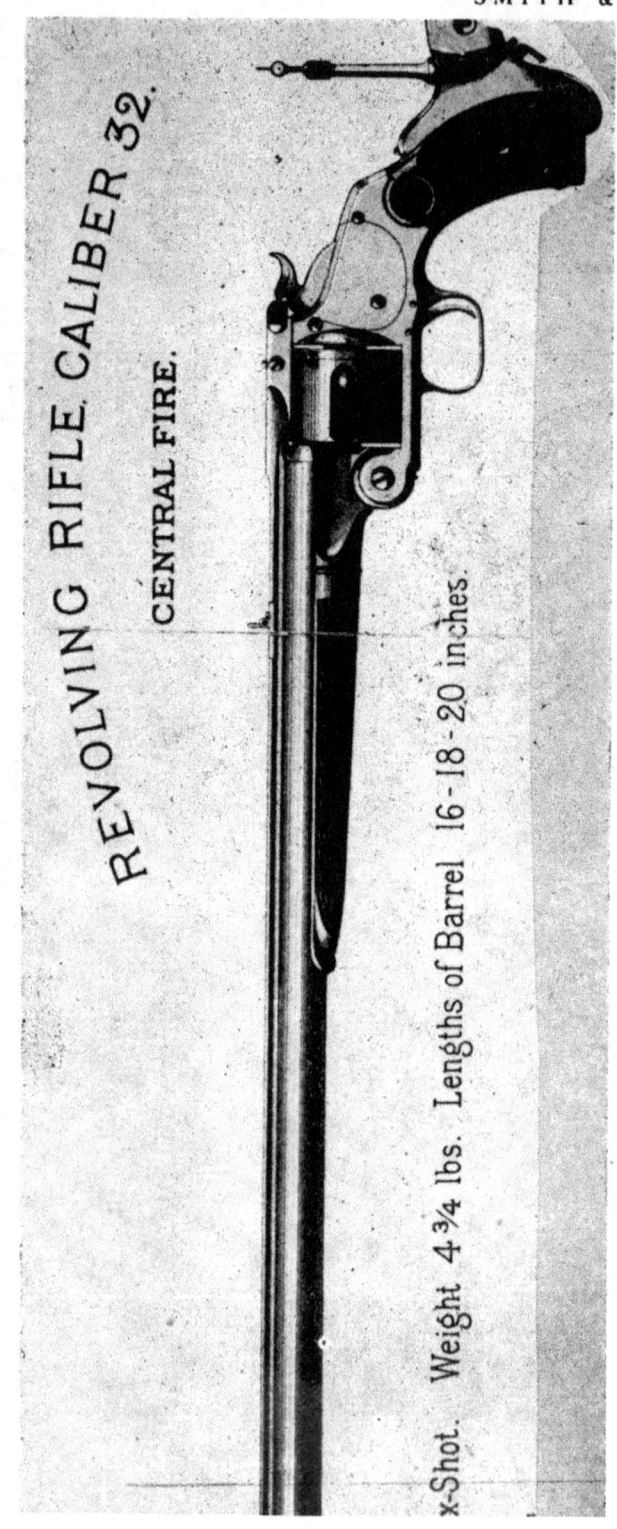

357 MAGNUM

HEAVY DUTY

38 M. & P.

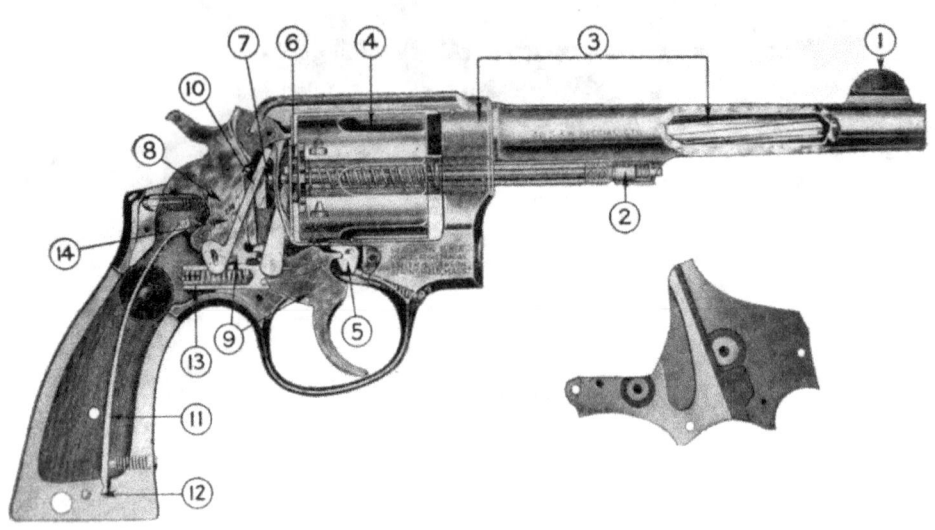

COMBAT MASTERPIECE

CENTENNIAL

MASTERPIECE

44 TARGET

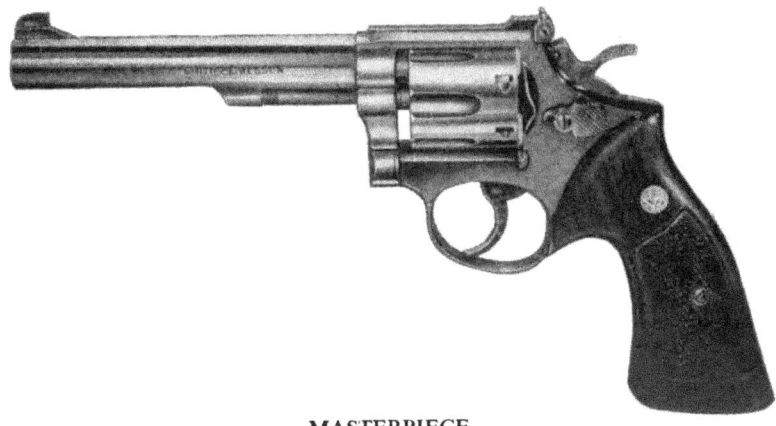

MASTERPIECE

SMITH & WESSON

MASTERPIECE

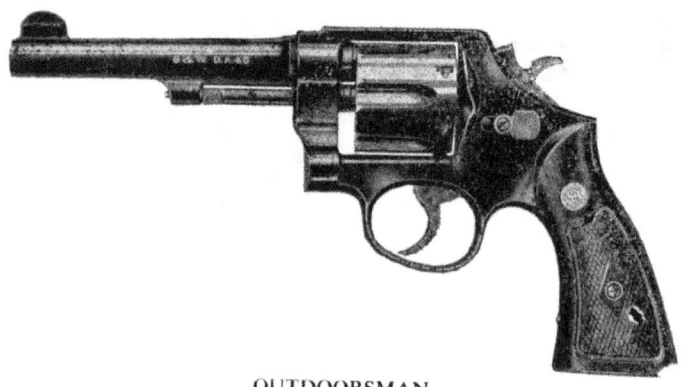

OUTDOORSMAN

KIT GUN

MASTERPIECE

HAND EJECTOR

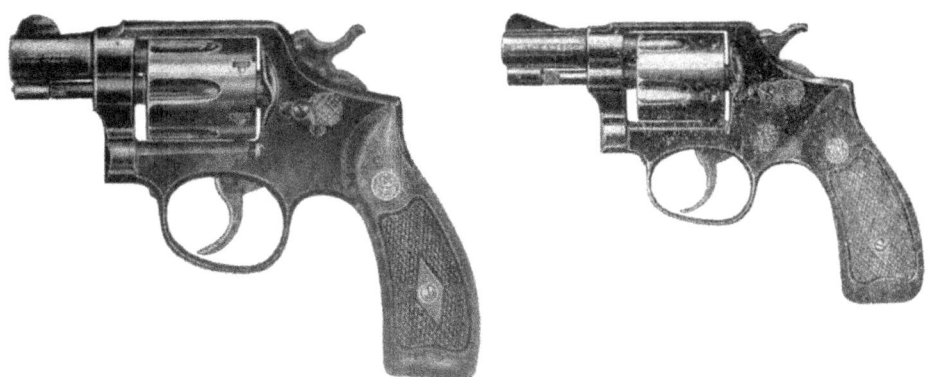
M. & P.

38 CHIEF'S SPECIAL

INDEX

A

A.C.P. cartridge, 65
Airweight revolvers, 66
Allen, E., 11, 46
Allen & Thurber, 11
American model, 37
Army model 1950, 65
Arrowsmith, G.A., 14
Automatic, 57

B

Bacon, Roger, 60
Ballistics, S. & W. cartridges, 93
Bekeart, P. B., 56
Bennett, Fred, 44
Bennett, W. W., 44
Berdan, Hiram, 37
Bull, H. W., 77
Bullet shapes, 27, 28

C

Centennial, 40, 66
Chattaway, J., 10
Chiefs Special, 65
Clement, C. P., 57
Cody "Buffalo Bill," 38
Colt, 8, 11, 26, 29, 34, 36, 49, 71
Current Prices, 84-91

D

Dodge, W. C., 37
Double action, 45
Double action perfected, 56
Dyer, A. B., 38

F

Fetterman massacre, 32
Flobert, 11, 12
Ford, H., 71
Forsyth, A., 28
Forty-five target, 65
Franklin, W. B., 36
Frontier, D. A., 48
Frontier, S. A., 47

G

Government stampings, 93
Gunpowder, 60, 61

H

Hamilton, A., 70
Hand Ejector, 55
Hanson & Golden, 30
Hartley, S. & G., 33
Heavy Duty 22/32, 56
Heavy Duty 38/44, 60
Hellstrom, C. R., 67, 72
Henry, B. T., 15, 16, 35
Hickok, "Wild Bill," 34, 74

Houllier, 29
Hunt, Walter, 13, 14, 30

I

Ignition systems, 27, 61
Industrial Revolution, 69

J

Jennings, L., 14

K

K-200, 64
Keith, E., 60
Kentucky rifle, 69, 70
King, C. A., 37
Kit Gun 22/32, 60

L

Leavitt revolver, 7, 8
LeFaucheux, 29
Leland, H. M., 71
Leonard Pistol, 9

M

M Hand Ejector, 51
Magnum, 60, 61, 62
Mark 2 Hand Ejector, 58
Mason, W., 43, 46, 49
Mass. Arms Co., 8
Masterpiece, 64
Maynard, E., 18, 28
Military & Police, 50, 63, 64
Minie, R., 27

N

New Departure, 48
New Haven Arms, 35
New Model 44, 48
New Model S. A. 44, 44

O

One, one-half old model, 34
Ordnance Test, 1876, 93
Outdoorsman, 59

P

Paine, I. A., 44, 45
Palmer, C. C., 12, 16, 22
Paris Int. Exhibition, 35
Pauley, 28, 29
Pepy's Diary, 29
Percussion Plains rifle, 70
Pettingill, C. S., 48

R

Regulation Police, 51

Renette, G., 44
Remington, 43, 60
Repeating rifle, 45
Robbins & Lawrence, 9, 15, 16
Russian model, 38

S

S. & W. models, 95
S. & W. peculiar points, 94
S. & W. 1st model, 26
Safety hammerless, 49
Schofield, G. W., 39, 40
Schurz, C., 69
Sharps, C., 28
Single action 38/44, 48
Single shot 1905, 51
Smokeless powder, 60, 61
Special Victory model, 64, 65
Spencer, C. M., 33
Springfield, 25, 41
Straight Line 22, 58

T

Target 44, 59
Terrier, 63
Third model, 49
Thuer, F. A., 36, 37
Thurber, C., 11, 70
Two-one model, 43

U

Union M. C. Co., 33

V

Volcanic, 20

W

Wesson, Daniel, 7, 9
  Douglas, 77
  Edwin, 7
  Frank H., 77
  Franklin, 9, 45
  Harold, 77, 79
  J. H., 53
  Joseph, 47, 77
  Rufus, 6
  Silas, 73
  Thomas, 7
  Victor, 77
  W. H., 47
  Walter, 76, 77
Wesson Favorite, 47
White, Rollin, 23, 24, 32, 36, 37, 74
Whitney, E., 11, 69, 71
Winchester, 21, 22, 50
Witsil, W. E., 60
Worcester, 5, 6, 70